The Quantum Dimensional Review of Newton and Einstein

Lawrence Dawson

The Paradigm Company, Boise Idaho

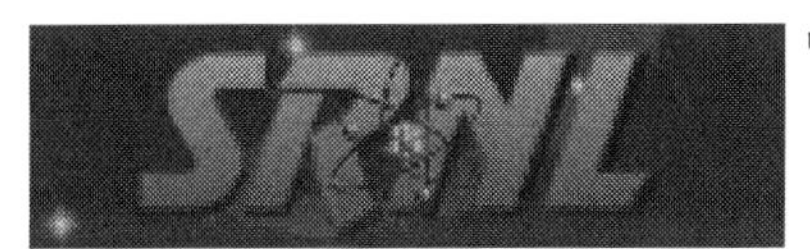

The Snake River N-Radiation Lab
www.srnrl.com
P.O. Box 311 Wilder, ID 83676

Reconciling Newton's and Einstein's Gravitational Curvatures of Space by Identifying the Exaggerated Horizon Moon or Sun as Caused by Quantum-Squared Vacuum Lensing

and the Derivation of Newton's Gravitational Constant from the Quantum-Dimensional Definition of Mass

Lawrence Dawson
The Snake River N-Radiation Lab

"Quantum-dimensional mathematics are becoming increasingly exact in their predictions and interpretations of physical events, increasingly complex in their mathematical detail, and increasingly alien to the belief systems governing contemporary science."
Lawrence Dawson

Introduction to Quantum-Dimensional Geometry and the Relationship between Quantum-Squared Vacuum and a Strict Euclidean Definition of Vacuum

The preliminary interfaces between the quantum and the Euclidean dimensions: the derivative of an Euclidean geometric unit of measure[1] reveals its quantum value[2] :

The derivative of an Euclidean unit of measure always establishes the quantum value of the unit of measure.

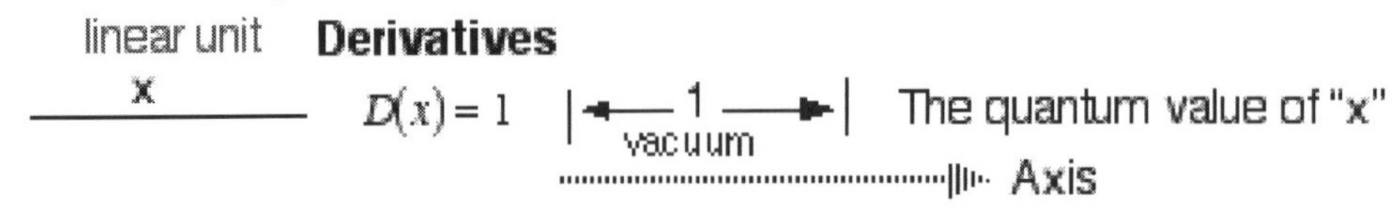

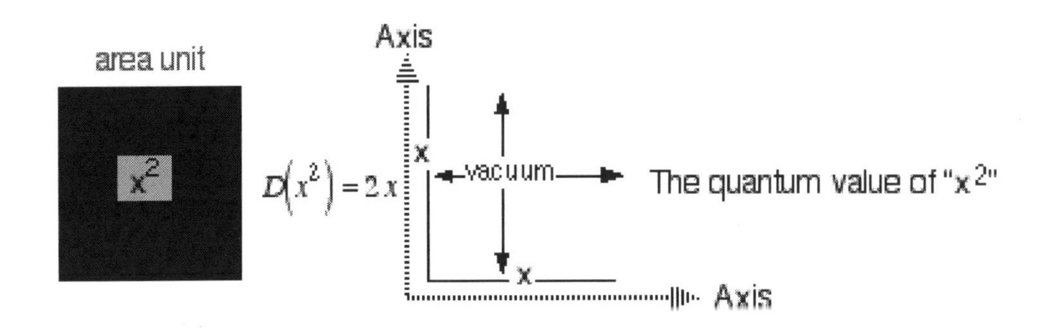

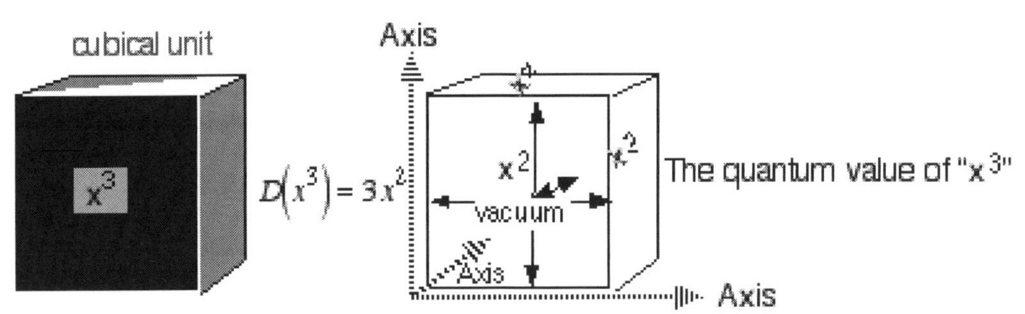

In the above illustration it can be seen that the calculus derivative of an Euclidean linear unit "x" is equal to "1" which is the quantum value since the distance is no longer divisible. That is, "x" is now defined as two end points with a non-divisible vacuum of separation which is the definition of a quantum. Similarly, the derivative of the Euclidean unit of area produces "2x," or two linear values of "x" which outline the unit of area as a quantum space (i.e. no longer divisible). Finally, the calculus derivative of the Euclidean cubical unit produces "3" area units of "x^2" which outline the cubical unit as a quantum space (no longer divisible and composed of vacuum). It is called a "*plenum.*"

Vacuum is the derivative of a three-dimensional unit of measure.

René Descartes identified a major problem in graphing the empty space surrounding a geometric solid and the sold itself using the same three Cartesian axii[3]. He recognized that the solid contained a continuum of points along any line contained within the solid. However, the surrounding vacuum must also contain a continuum of points if a continuum of distance measures are to be made within the vacuum. He concluded that vacuum itself must be a solid which he designated as "aether." Descartes' "vacuum as a solid" solution failed to recognize that it is the existence of a separate quantum dimension which distinguishes vacuum from any solid; that the surrounding vacuum must

[1] In an "Euclidean unit of measure" all the lines contained within it provide a continuum of points.

[2] A quantum value supplies indivisible space which does not contain lines providing a continuum of points.

[3] See *The Quantum Dimension;* Lawrence Dawson, Paradigm Publishing; 2009; ISBN: 0-941995-24-0. *"The Concept of the Quantum as a Separate, Fourth Geometric Dimension",* Page 202

be composed of quantum units, not as a continuum of points . Quantum-defined vacuum simply cannot be graphed using Cartesian coordinates since the quantum cannot be constrained by the Cartesian graph's axii.

> *"The quantum dimension cannot be graphed upon the standard Cartesian graph with the "y" axis representing the quantum dimension and the "x" axis representing an intersecting Euclidean dimension.*
>
> *a. The "y" distance cannot be constrained to the quantum distance by a Cartesian graph. All points along "x" at "y" will make independent quantums with all points along "y+1" and "y-1." The quantum distance value will not be consistent and cannot be made consistent.* [4] "

The Quantum must be Graphed as the Unit of Vacuum Produced by a Single Quantum Point Opposing an Euclidean Line of a Distance Equal to the Quantum; the Euclidean Line being "Kinked[5]" into Curvature to Produce Vacuous Volume

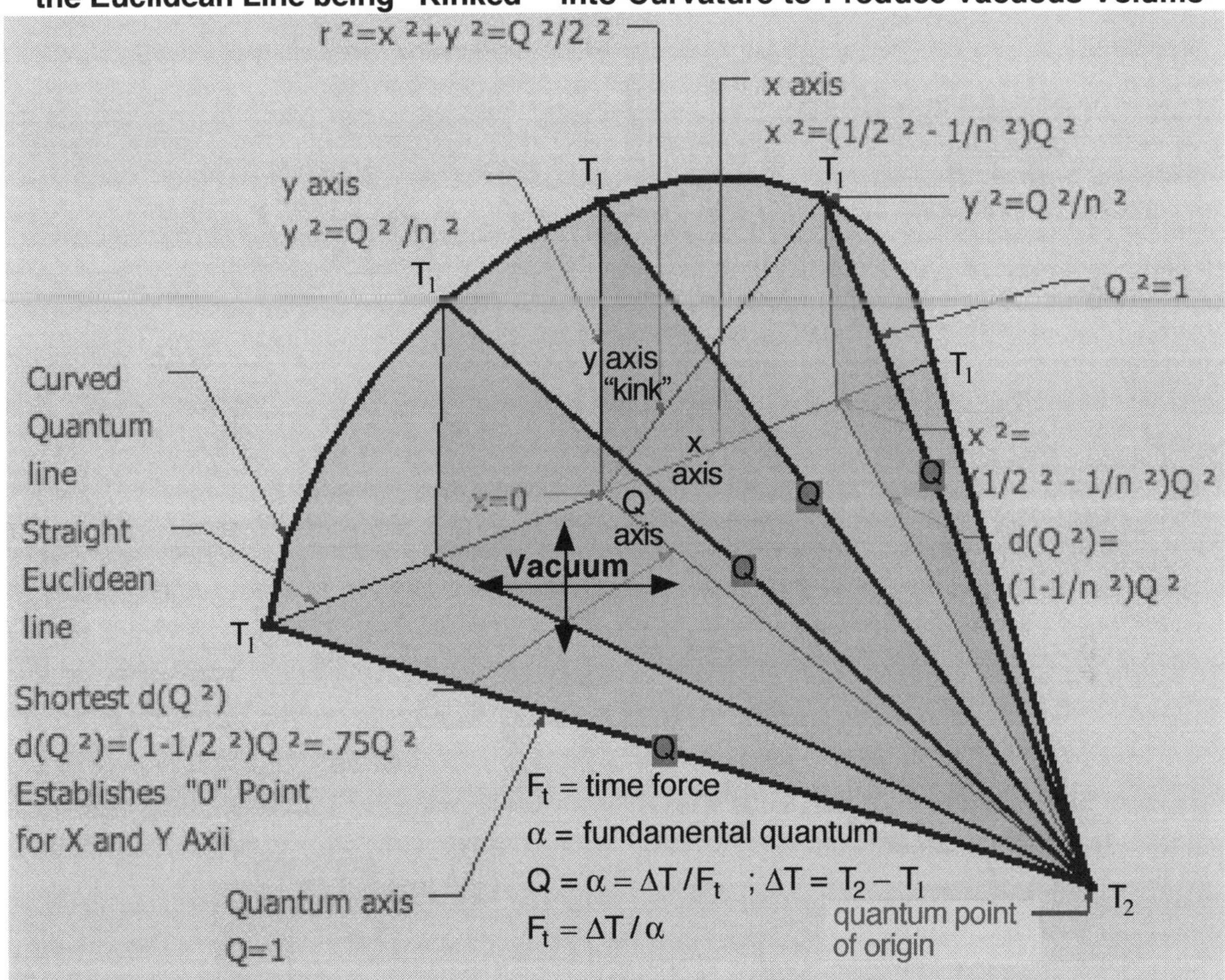

An Euclidean line opposing a quantum point is "kinked " into curvature by quantum time-force (produced by a potential time factor) to retain integrity of quantum distance. Thus, the quantum-squared produces a unit of volume. Vacuum is actually two-dimensional quantum, not three-dimensional Euclidean.

By preliminary quantum geometry, vacuum is the derivative of the Euclidean solid. This derivative renders three planes defined by all possible two-axis combinations from a three axis Cartesian graph. The volume contained between these planes is vacuum. The graph composes what is termed a "*plenum.*" However, the point of origin for the Cartesian "plenum" graph is arbitrary and

---

[4] *Lecture Notes Quantum Geometry;* http://www.srnrl.com/id12.html

[5] "See the "1+1 dimensional kink" in *Solitons* at *physics.usc.edu/~vongehr/solitons_html/solitons.pdf*

can take any point within the vacuum. The imprecision of the point of origin provides a continuum of possible planes defining the vacuum. This ambiguity of the "*plenum*" graph's point of origin, makes the measured vacuum to be the solid that Descartes argued for it. The quantum dimensional graph provides that any point in space can be the quantum point of origin to establish an exact quantum value of vacuum without producing such a continuum of planes. The three planes of the Cartesian "plenum" graph are replaced by a single "kinked" plane as the base of a cone with vacuum being defined by the volume of the cone.

The quantum derivative of a theoretical Euclidean unit of vacuum converts to a function of the "kink squared" for the actual quantum-squared unit of vacuum. The three planes arrayed along three Cartesian axii becomes the denominator factor for the volume of a cone as defined by the single "kinked" Euclidean plane:

$$Vacuum = D\left(x^3\right) = 3x^2;\ x^2 = kink^2_{d(Q_{n=2})}\pi\left[d\left(Q_{n=2}\right) = \sqrt{(1-1/2)Q^2}\right]/3$$

The "3" Cartesian axii are converted to the "3" denominator from the volume formula for the quantum-defined vacuum cone. Quantum vacuum is two-dimensional "kinked" volume.

VACUUM AS A CARTESIAN SOLID: Its alleged application to gravity by Descartes and Einsteinian General Relativity

At the dawn of scientific geometry, its founder René Descartes proposed that if vacuum is to be graphed on the same three Cartesian axii as a solid, then vacuum must also be a mysterious solid which Descartes later came to call "aether." As we have proved above, Descartes is right if vacuum is defined as the derivative of a solid; as the empty space between the planes made by all two axii combinations of the three axii graph. This produces a "*plenum*" graph and the point of origin for such a graph is free floating. In that case, the planes containing the empty space become continuums establishing vacuum as a Cartesian solid. The unrecognized source of the Cartesian solid is the floating point of origin. Newton would later reject this notion of a Cartesian solid vacuum as incompatible with the laws of motion.

> *"One of the more controversial positions [Descartes'] Principles forwarded, at least according to Newton, was that a vacuum was impossible. Descartes' rejection of the possibility of a vacuum followed from his commitment to the view that the essence of body was extension. Given that extension is an attribute, and that nothing cannot possess any attributes (AT VIIIA 25; CSM I 210), it follows that "nothingness cannot possess any extension" (AT VIIIA 50; CSM I 231). So, any instance of extension would entail the presence of some substance (AT VIIIA 25; CSM I 210). In other words, vacuum, taken as an extended nothing, is a flat contradiction. The corporeal universe is thus a plenum, individual bodies separated only by their surfaces. Newton argued in his De Gravitatione and Principia that the concept of motion becomes problematic if the universe is taken to be a plenum."* [6]

Newton's objection to a vacuum as a Cartesian solid was that "aether drag" would interfere with th constancy of motion. Inertia holds that a body in motion tends to remain in motion, but vacuum resistance would possibly decelerate a body after its forceful acceleration to a constant velocity.

Actually, Descartes theory of vacuum geometry had a logical problem from its inception. His premise that "nothing" is the equivalent of vacuum is incorrect. The actual equivalent of vacuum is "emptiness," not "nothingness," and emptiness, or the lack of geometric points[7], can contain a "field of force." "Extension" may be the result of force separating two quantum points. Geometric points have no dimension, only position. Therefore, Those points constitute "nothing" in Descartes' vocabulary, even though they have "position" from which an extension can be measured. Working before the concept of "fields" had been developed by Newton, Descartes could not recognize that an extension from a point of origin of fixed position was possible for a field.

From his conception of vacuum as a Cartesian solid, Descartes derived a mechanical explanation

[6] Smith, Kurt, "Descartes' Life and Works", The Stanford Encyclopedia of Philosophy (Fall 2012 Edition), Edward N. Zalta (ed.), URL = <http://plato.stanford.edu/archives/fall2012/entries/descartes-works/>.

[7] The quantum definition.

for gravity:

> *"René Descartes proposed in 1644 that no empty space can exist and that space must consequently be filled with matter. The parts of this matter tend to move in straight paths, but because they lie close together, they can't move freely, which according to Descartes implies that every motion is circular, so the aether is filled with vortices....... According to Descartes, this inward pressure [of vortices] is nothing else than gravity."*[8]

Descartes explanation of gravity as vortices within a Cartesian solid ultimately fell before the more rigorous mathematics of Newton's gravitational mechanics. However, the idea of gravity as a function of a Cartesian solid was to be resurrected by Albert Einstein in general relativity.

Einstein Resurrects Vacuum as a Cartesian Solid in his General Relativity Equations

Almost three centuries later, Albert Einstein would resurrect the idea of vacuum as a Cartesian solid to make it the foundation of his gravitational field equations for general relativity:.

> *"Descartes argued somewhat on these lines: space is identical with extension, but extension is connected with bodies; thus there is no space without bodies and hence no empty space."*[9]

In order to avoid Newton's objection to "aether drag" as incompatible with the laws governing the constancy of motion, Einstein proposed an alternative formulation for the Cartesian solid. Instead of a mass moving through the Cartesian solid finding resistance, he proposed that the mass was figuratively "riding on top of the Cartesian solid." By depressing or "curving" the Cartesian solid out of its way, the mass removed all resistance to its motion.

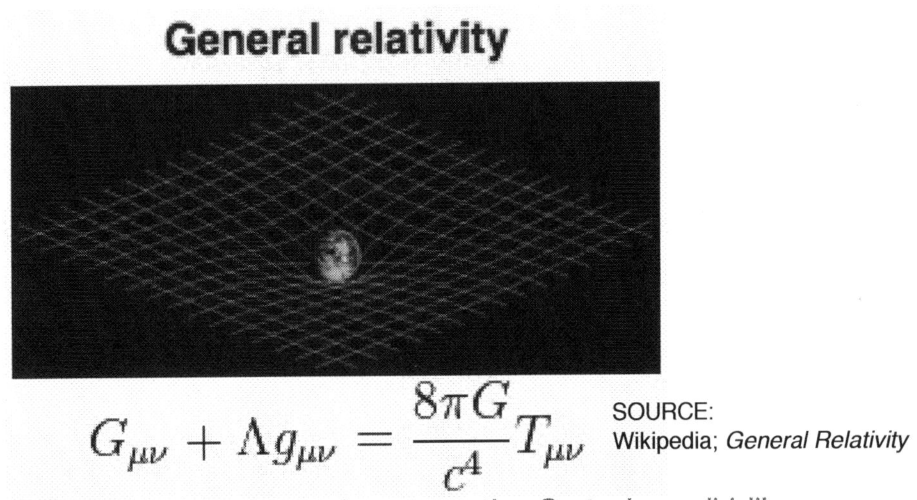

SOURCE:
Wikipedia; *General Relativity*

By general relativity, mass depresses the *Cartesian solid* like a heavy ball on a flexible membrane. This distortion of the *Cartesian solid* is alleged to remove any resistance to its motion.

Albert Einstein's general relativity equations are retrograde theory which harkens back to the discredited *plenum* solid vacuum of Rene Descartes. Even if he resolved the "aether drag" issue with his "space curvature tensor" for the Cartesian solid, Einstein's retrograde formulations were still inadequate. Einstein's equations have no autonomous geometric and derive all geometrics from Newton's Gravitational constant rather than deriving Newton's Gravitational constant from a set of autonomous geometrics.

> *"The Einstein field equations [EFE] are used to determine the spacetime geometry resulting from the presence of mass-energy and linear momentum, that is, they determine the metric tensor of spacetime for a given arrangement of stress–energy in*

---

[8] From Wikipedia. *"Mechanical explanations of gravitation; Vortex"* *http://en.wikipedia.org/wiki/Mechanical_explanations_of_gravitation*

[9] Einstein, Albert *Relativity* ; Three Rivers Press, ISBN 0-517-88441-0 ; p. 156

the spacetime. The relationship between the metric tensor and the Einstein tensor allows the EFE to be written as a set of nonlinear partial differential equations when used in this way..........the EFE reduces to Newton's law of gravitation where the gravitational field is weak and velocities are much less than the speed of light[10] ."

The Einsteinian nonlinear gravitational equation is written in the following form with the symbolism defined:

$$G_{\mu\upsilon} + \Lambda g_{\mu\upsilon} = \frac{8\pi G}{c^4} T_{\mu\upsilon}$$

Einstein's Cosmological Constant: "The constant has the effect of a repulsive force [attached to vacuum]....." ENCYCLOPEDIA BRITANNICA

$(G_{\mu\upsilon} = \text{Einstein's curvature tensor})$; $(\Lambda = \text{Einstein's cosmological constant})$

$(g_{\mu\upsilon} = \text{Einstein's metric tensor})$; $(T_{\mu\upsilon} = \text{Einstein's stress energy tensor.})$

$(G = \text{Newton's gravitational constant})$; $(c = \text{speed of light in a vacuum})$

The first thing to understand about the Einsteinian gravitational formulations is that they are seldom, if ever, predictive:

> *"It is important to realize that the Einstein field equations alone are not enough to determine the evolution of a gravitational system in many cases. They depend on the stress-energy tensor, which depends on the dynamics of matter and energy (such as trajectories of moving particles), which in turn depends on the gravitational field. If one is only interested in the weak field limit of the theory, the dynamics of matter can be computed using special relativity methods and/or Newtonian laws of gravity and then placing the resulting stress-energy tensor into the Einstein field equations. But if the exact solution is required or a solution describing strong fields, the evolution of the metric and the stress-energy tensor must be solved for together."*[11]

On the left side of the equation are the geometrics for Einstein's self-identified version of a Cartesian solid vacuum. These two tensors (tensors for curvature and for metric conversion) determine how much curvature the Cartesian solid will undergo when put under "stress" by Newtonian defined gravity on the right hand side of the equation *times* a "stress energy tensor."

The gravitational factor on the right side is modified by a variable "stress energy tensor" which identifies further enforced curvature of the Cartesian solid vacuum by motion (hence eliminating possible "Aether drag"). The exact amount of alleged curvature of the Cartesian solid vacuum can never be exactly predicted.

If the Einsteinian retrograde formula does not have predictability capacity, what is its purpose? It predicts the existence of a curved Cartesian solid vacuum in the presence of mass and thus, it is said, identifies many modern phenomenon such as "gravitational lensing" which have recently been observed[12]. Even though the Einsteinian gravitational field equations have been proved correct with respect to the curvature of mass, they are still deficient geometrically, a deficiency which can be corrected by a review of the proofs for Einsteinian mass induced spacial curvature. ***It can be proved that the mass-induced curvature of vacuum originates from fixed quantum points of origin, not the floating points of Einstein's Cartesian Solid Vacuum.***

The Quantum-Dimensional View of an Object on a Gravitational Horizon is the Exact Equivalent of Einstein's Predicted Geometric Curvature of Vacuum by Mass

For well over a hundred years psychologists have attempted to explain the fact that we see a full

[10] Carroll, Sean (2004). Spacetime and Geometry - An Introduction to General Relativity. pp. 151–159. ISBN 0-8053-8732-3.

[11] *"Solutions of the Einstein field equations";* Wikipedia. http://en.wikipedia.org/wiki/Solutions_of_the_Einstein_field_equations

[12] http://en.wikipedia.org/wiki/General_relativity

moon on the horizon as twice the size of a zenith moon. In this time period, no psychological theory of human perceptual error has been able to establish the phenomenon as purely "illusionary[13] ." The "illusionary" nature of the horizon moon is assumed because a single lens producing a "flat" photograph, eliminates the apparent size variations between the horizonal moon and the zenith moon. Psychologists could not recognize that the stereoscopic view from a set of human eyes provides three dimensional information while the single lens view of a camera does not. Perceptual psychology did not have quantum-dimensional optics and, therefore, could not recognize the phenomenon as an actual physical event.

The real explanation is that the horizonal moon provides a "lens bias" which is only perceptible using a stereo or three-dimensional view. The existence of a *focal bias* for a full moon viewed on the horizon is a mathematical certainty without resorting to quantum-dimensional mathematics.

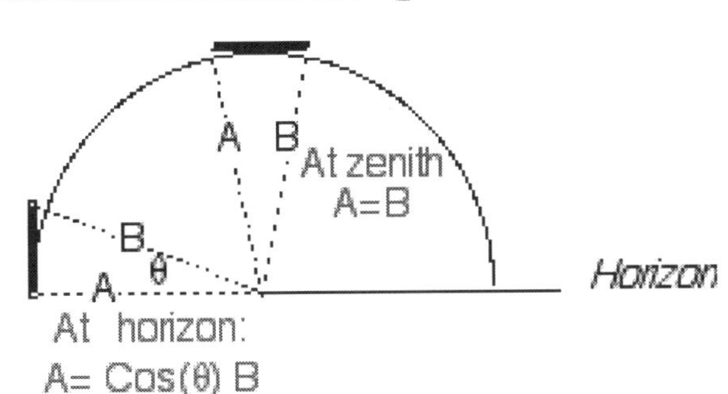

HYPOTHESIS: The stereoscopic, three-dimensional view across the gravitational horizon toward an object which is in gravitational conjunction with the body of viewpoint reveals the curvature imposed by the object upon quantum space.

A focal bias exists on the horizon because the line-of-sight ,"B," from the focal point to the upper limb of the moon is longer than the line-of-site, "A" from the focal point to the lower limb. This bias is not true for the zenith at which lines of sight to the focal point are equal.

However, this linear focal bias is much too small to explain the increase actually observed (twice the size of the original). It has been rejected by psychologists as the explanation since the angle "theta" is only "0.5°."[14] This is much too small to supply a perceptible difference between distances "A" and "B." This changes dramatically when the bias is viewed stereoscopically using the spacial model supplied by quantum-dimensional geometry. The curvature of space lenses a significantly biased view of the moon and one which fits the actual data.

The Lensing Bias of an Object Viewed Stereoscopically on the Gravitational Horizon of a Mass is its Quantum-Dimensional View

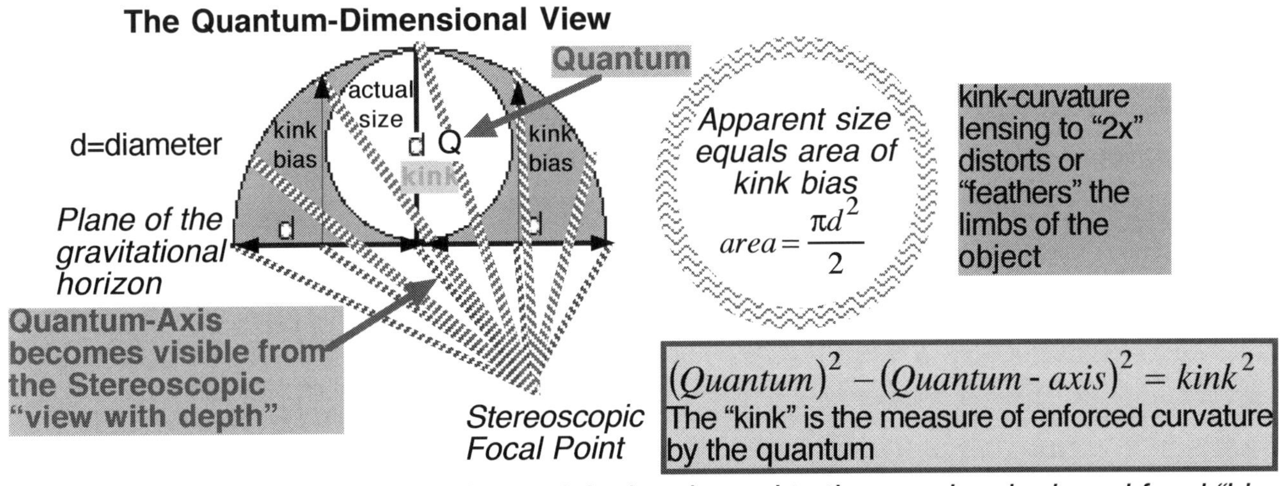

Apparent area is twice the size of the original and equal to the area in a horizonal focal "bias field" which is lensed by "kink curvature" of space and which is discernible only with a stereoscopic view. .

---

[13] *"The Moon Illusion Explained"*; Don McCready ; Psychology Department, University of Wisconsin-Whitewater. *http://facstaff.uww.edu/mccreadd/intro9.htm.*

[14] McCready. Op. cit.

Viewed stereoscopically, the horizonal focal bias is converted from the linear bias to a field bias with the height of the object converted to the radius of the field. This produces a field bias which has twice the area of the object and produces an apparent size equal to the area of the bias field. The angle of bias is no longer relevant as long as it is sufficient to produce a discernible area in the celestial canopy. The horizonal focal bias field reproduces the quantum squared view.

The Lensing Bias of a view across the gravitational horizon is the exact equivalent of the geometric biasing of vacuum as predicted by Einstein's Gravitational Field Equations

Image with a stereoscopically viewed horizonal focal bias [15]

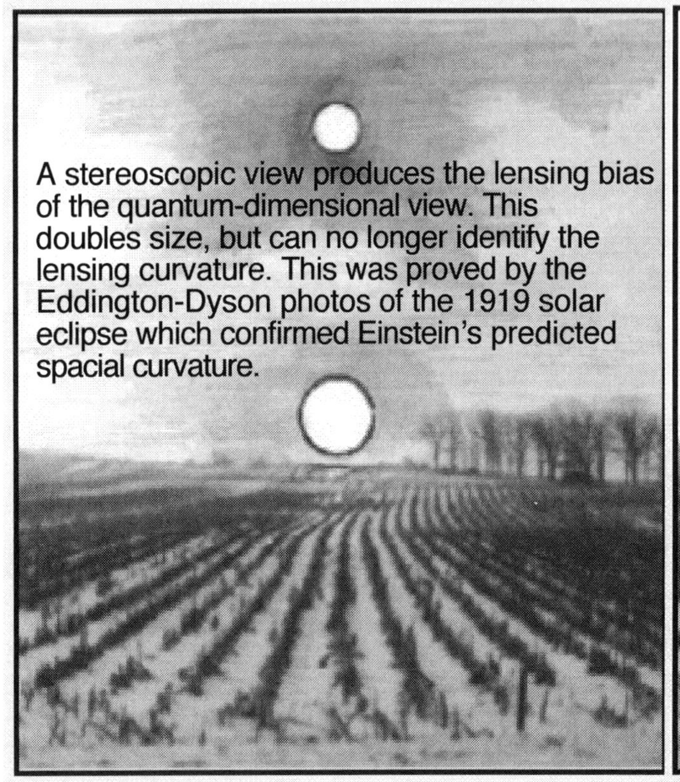

Image from a single photo lens does not detect horizonal focal bias [15]

A single photo-lens cannot detect the quantum-dimensional view *(this is an actual double exposure)*. However,the "flat view" can detect the spacial curvature by the mass. This was also proved by the 1919 Eddington-Dyson eclipse photos which confirmed Einstein's prediction.

The increase in perceived size of the horizonal moon is not an optical illusion but an actual lensing of the light by mass-induced curved space. This increase in size by bending light through quantum curved space must be viewed along three axii defining the plane of the horizon, not two axii. The three axii stereoscopic view produces a quantum squared aperture. This quantum lensing of the moon's light operates by quantum optical principles which are not equivalent to those governing Euclidean defined lenses. This lensing of the light through quantum curvature of space is no more an optical illusion than is the apparent increase in size of the same moon when viewed through a telescopic lens.

The curvature of space by the mass cannot be detected while light is being lensed by that curvature to produce an apparent increase in size. This requires no mystical principle to understand, but is commonly observed with telescopic lenses. The curvature of the lens cannot be detected while viewing light through them. The curvature of the lens can only be detected when viewing the lens as an independent object and while they are not functioning to curve the light which is falling upon the eye.

This inability of the mass-induced curvature of space to be detected while it lenses light has been proved by existing data. That proof has lain unrecognized in the 1919 solar eclipse data by which Eddington and Dyson measured the curvature of space in support of Einstein's prediction. The amount of mass induced curvature predicted by Einstein is found to exactly equal the lensing bias

[15] McCready. Op. cit.

of that curved space. Both are empirically confirmed by revisiting the Eddington/Dyson data.

The Eddington-Dyson Solar Eclipse Photos Prove both Einstein's Prediction of Mass Curvature of Vacuum and its Relationship to Quantum-Dimensional Lensing Bias

> *"One of the most famous measurements in the history of 20th-century astronomy was made over the course of several months in 1919. Teams of observers from the Greenwich and Cambridge observatories in the UK traveled to Brazil and western Africa to observe a total solar eclipse that took place on 29 May 1919. Their aim was to establish whether the paths of light rays were deflected in passing through the strong gravitational field of the Sun. Their observations were subsequently presented as establishing the soundness of general relativity; that is, the observations were more consistent with the predictions of the new gravitational theory developed by Albert Einstein than with the traditional Newtonian theory."*[16]

The Newtonian gravitational formulations predicts a curvature of space proximate to matter. In 1804 the German astronomer Johann Georg von Soldner had predicted a gravitational lensing of light by the sun's mass using strictly Newtonian gravitational mechanics. He predicted light would be bent by approximately 0.87 arc seconds[17]. Einstein's stress curvature of vacuum from his relativity field equations predicted a spacial curvature of double this amount (stress curvature of 1.75 arc seconds).

> *"In 1916, after he had developed the final version of his theory of general relativity, Einstein realized that there was an additional component to the light-deflection effect caused by the way that the Sun's mass curves spacetime around itself. Thus a straight path, or geodesic, near the Sun is curved, compared with a path through flat space. The extra deflection caused by that curvature is comparable to the deflection due solely to falling*[to gravitational influence]*, so that the general relativistic prediction calls for twice as great a shift in stellar positions— about 1.75" at the limb of the Sun— as does the Newtonian theory.*[5]
>
> [5] *A. Einstein, Ann. Phys. (Leipzig)* ***49****, 769 (1916).*
>
> *"As early as 1913, Einstein wrote to leading astronomers, trying to interest them in making a measurement of the effect he had predicted. Stars are not normally visible close to the Sun, though, a problem that required astronomers to take pictures of a field of stars around the Sun during a total solar eclipse."* [18]

Einstein's prediction of stress curvature of vacuum as twice that of gravitational curvature was tested by Cambridge University's A.S. Eddington who traveled to the island of Principe in equatorial Africa and by a team sent by Greenwich's F.W. Dyson to Sobral station in equatorial Brazil. Both teams photographed the 1919 solar eclipse against a backdrop of the Hyades star cluster. which was close to the Sun during the solar eclipse. The position of cluster stars photographed through the eclipse's corona light field, as compared to those cluster stars photographed in the night sky in the absence of the Sun, revealed how many arc seconds those stars had been displaced by the Sun's presence.

Eddington's eclipse plates were made near midday when the sun was near its zenith and therefore without horizonal bias. He used a set of night plates for the Hyades cluster which had been made in Oxford. He hadn't want to wait the 5 to 6 months for an equivalent night view of the cluster from Africa. By testing night sky clusters photographed in Oxford and Africa he was able to eliminate positional bias between the eclipse view of the Hyades cluster from Africa and the night view from Oxford. With his Oxford night view plates in hand, Eddington immediately calculated a curvature closer to the predicted Einsteinian stress curvature, rather than the strictly gravitational curvature (Eddington: 1.60 arc seconds ±0.3 arc seconds or a maximum of 1.9 arc seconds[19]).

---

[16] Kennefick, Daniel *"Testing relativity from the 1919 eclipse— a question of bias."* Physics Today, March 2009

[17] J. G. von Soldner, Berl. Astron. Jahrb., 161 (1804).

[18] Kennefick, Op. cit. p.38

[19] *"The Eclipse of 1919 May 29 and the Theory of Relativity."* Monthly Notices of the Royal Astronomical Society, Feb. 1920. Vol. 80, p.415

The second set of eclipse plates were made at approximately 7:30 in the morning with the eclipsed sun near the horizon at 22° azimuth[20]. The eclipse was photographed with two telescopes, a 19 inch multi-element wide angle astrographic lens[21] and a single element 4 inch lens. The multi-elements which compose a wide-angle astrographic lens are capable of a stereoscopic or three dimensional view. The single element of the 4 inch lens gave a "flat" view. This was confirmed when the astrographic lens gave the Hyades stars which were visible through the eclipse's corona light field the characteristic "feathering" of limb details for the quantum-dimensional lensing of light through Einstein's mass induced curvature of space.

> *"In the immediate aftermath of the eclipse, onsite development of some plates alerted Crommelin and Davidson that the astrographic setup had lost focus during the eclipse. The* [Hyades cluster] *stars* [visible through the corona] *were noticeably streaky, a problem reported by Dyson at a meeting of the Royal Astronomical Society as early as 13 June. Disturbingly, when the comparison plates were taken two months later, the instrument was once again in focus.*[22] "

Further, the lensing through the Earth's mass-induced curvature (when viewed on the horizon) will eliminate the view of the mass-induced curvature around the Sun's horizon. Even while being blurred by incorporation into the focal bias of the Sun's coronas light field, the stars positions are constant relative to the Sun due to a stereoscopic infinity focus which the Sun does not share. Due to this stereoscopic focal difference, the positions of the stars relative to the night view will remain constant while the position of the Sun will not.

The difference in stereoscopic focus under curved-space lensing does not cause the exaggerated sun to "cover" the peripheral stars, only to move closer to them. This is the actual explanation of the "blurring" of the stars within the coronas light field which was discovered in the Brazilian horizonal astrographic photos. This contraction of distances around the solar periphery predicts a variation in the bending of starlight outward by the Sun[23] . The biased view under curved-space horizonal lensing will give exactly one half the bending of the unbiased or "flat" view. This is true because the stereoscopic view of the horizonal image of the sun lenses it to twice its size relative to the infinity-focused positions of the peripheral stars. Einstein's predicted mass-induced spacial curvature bending will be eliminated remaindering only gravitational bending.

The Sobral, Brazil horizonal photos taken of the eclipse with the stereoscopically-focused astrographic lens produced a bending value which was approximately one-half that measured by Eddington's "zenith" astrographic photos of the eclipse. The bending value was also approximately one-half that measured by the "flat" four-inch lens used at Sobral.

The Record of the Arc Seconds of Curvature from the 1919 Solar Eclipse Data[24]

$$\{\text{Eddington's astrographic - lens photo at zenith}\} = 1.6 \pm 0.3 \text{ arc seconds} \quad \langle \textbf{non - biased view} \rangle$$

$$\{\text{Sobral single element lens photo at horizon}\} = 1.98 \pm 0.12 \text{ arc seconds} \quad \langle \textbf{non - biased view} \rangle$$

$$\{\text{Sobral astrographic - lens photo at horizon}\} = 0.93 \pm 0.3 \text{ arc seconds} \quad \langle \textbf{biased view} \rangle$$

PREDICTED: $2(\text{gravity curvature}) = (\text{mass induced curvature}) + (\text{gravity curvature})$

$$2(0.87 \text{ arc seconds}) = (1.74 + 1) \text{ arc seconds}$$

EXPERIMENTAL: $2(0.93") = 1.86" = (1.6 + 0.26)" = (1.98 - 0.12)"$

Despite raging headlines to the contrary, the Eddington-Dyson photographs of the 1919 solar eclipse never "overthrew" Newtonian gravitational mechanics by Einstein's general relativity.

---

[20] Calculated from *Royal Astronomical Society* data given in their report *"The Eclipse of 1919 May 29 and the Theory of Relativity."* Op. cit.

[21] Ibid.

[22] Kennefick, Daniel "*Testing relativity from the 1919 eclipse— a question of bias.*" Physics Today, March 2009. P. 41

[23] It is predicted that peripheral starlight will be bent outward relative to the Sun by both gravity and Einstein's mass-induced curvature of space.

[24] Data given in Feb. 1920 *Royal Astronomical Society* report. Op. cit.

Einstein never contested the fact that Newtonian gravity could bend light. He simply asserted that his mass-induced stress tensors, as applied to vacuum, would multiply the Newtonian gravitational curvature of vacuum by "2x." Gravitational curvature bias is "0" when viewed *from within the gravitational horizon* — which includes any viewpoint from the surface of the earth. However, gravitational curvature bias is not "0" when viewed *across the gravitational horizon* as with our view across the sun's periphery during the solar eclipse. It is the gravitational curvature bias viewed across the gravitational horizon which is being multiplied by "2x" by the general relativity stress tensors.

We have demonstrated mathematically and empirically that a stereoscopic horizonal view of a rising mass of perceptible sky area will be lensed to a perceived value of "2x" by quantum-dimensional curvature of space. It is lensed to Einstein's predicted value for tensor-produced spacial curvature. Thus, the stereoscopic horizonal view of the rising mass eliminates Einstein's tensor-induced curvature and remainders only gravitational curvature.

For the photographs of the 1919 eclipse, the astrographic lens used on the horizonal view in Brazil was a stereoscopic viewpoint which suffered the predicted lensing bias. This bias was indicated by the "feathering" of the Hyades Cluster stars viewed through the light-field of the eclipse corona, a feature which appeared in no other photograph. Further the measurement of curvature by the Sobral horizonal view using the astrographic lens produced the gravitational curvature value, not Einstein's tensor-induced curvature value. The Einsteinian curvature had been eliminated by horizonal lensing bias from the astrographic lens. Both "non biased" views of the eclipse produced measurements of Einstein's predicted tensor-induced curvature. Both were twice the "biased" value of the astrographic lens, when calculated within the error ranges reported by the researchers.

Horiz.Astro.	**Zenith Astro.**	**Horiz. Flat**
$2(0.93'') = 1.86''$	$= (1.6 + 0.26)''$	$= (1.98 - 0.12)''$
	error range: ± 0.3	*error range*: ± 0.12

The Eddington-Dyson photographs of the 1919 solar eclipse supported Einstein's mass induced vacuum tensor and the general relativity field equations. Hidden in the data for nearly a hundred years, however, was a second discovery of equal importance. The photographs— which were once believed to present the greatest difficulty for the research— were the Brazilian astrographic lens photos with their distorted index stars as viewed through the eclipse corona and a "gravitational only" displacement which was at odds with the "Einstein supporting" values from other photographs.

These "problem" photos may yet prove to be the most important taken by the Eddington-Dyson team. They establish that Einstein's curved space is actually quantum squared and not the Cartesian solid vacuum he had believed it to be. The photos establish that Einstein's tensor-curved space can actually lens light along a gravitational horizon and at the rate which his field equation predicted. When viewed through this horizonal lensing bias, the Sun's tensor-curvature disappears, remaindering only gravitational curvature. If Einstein's tensor-curved space can become a light lens[25] and that lens is shown to be the quantum squared, then the originating curved space must also be the quantum-squared. This is the important discovery recently found within the Eddington-Dyson data, a discovery made 60 to 70 years after the deaths of the principles involved. Eddington-Dyson had also proved that vacuum is quantum, not a Cartesian solid, but the discovery could not be recognized because the period lacked the quantum-dimensional mathematics required to understand the "aberrant" Sobral astrographic photos.

The Revelation that Vacuum is Quantum-Squared Provides a Rational Einsteinian Cosmological Constant which reconciles with Newtonian Motion Mechanics

If Einstein's tensions-curved vacuum is the quantum squared, we can reconcile Einstein with Newtonian motion mechanics This reconciliation provides a seamless integration of the general relativity field equations with Newton's gravitational mathematics. Proof that Einstein's tension-curved space is quantum does rest only upon the capture of quantum-squared lensing bias by the

[25] The lensing must be by the tensor-curved space, not gravitational curved space because gravitational curvature from within the gravitational horizon is always "0." Gravitational curvature can only be detected by a view outside the gravitational horizon; that is, off the surface and looking back across the horizon of the now-external mass.

horizonal photography of the 1919 eclipse. It is also proved by the fact that quantum squared vacuum give this rational, Newtonian mathematical value to Einstein's "cosmological constant."

The cosmological constant is defined as "a repulsive force attached to vacum" or a tension energy in the absence of matter. Although the cosmological constant was a factor in the field equations, as confirmed by Eddington-Dyson in 1919, Einstein was to renounce it 10 years later under the influence of Edwin Hubble's "expanding universe" interpretation of his 1929 data table comparing stellar distance to measured redshift . Einstein's renunciation was premature. Quantum-dimensional mathematics have completely discounted Hubble's original "doppler effect" explanation[26]; the explanation which made the "static" cosmological constant seem impossible. .

Einstein's cosmological constant is the force of a "reserved time" which produces quantum-squared vacuum. Vacuum is not the "spacetime continuum" of Einstein's Cartesian solid. Rather, vacuum is quantum composed of both a distance quantum and a reserved-time quantum. The fundamental distance quantum is the the "alpha space" which has been measured in atomic structure as equal to "0.50216243346e-15 meters[27] ." The time quantum is the reserved time across the alpha space which is equal to "*1.6750335776e-24 seconds of reserved time*"

The Alpha Space and the Reserved Time Quantum

$$\Delta T = \langle \text{reserved time across alpha space} \rangle; \qquad c = \alpha / \Delta T$$

$$\alpha = (\text{alpha - space}) = 0.50216243346e\text{-}15 \; meters; \;\; \Delta T = 1.6750335776e\text{-}24\,\text{sec}.$$

$$F_t^2 = (\text{time - force})^2 = \frac{\Delta T^2}{\alpha^2} = \frac{1}{c^2}; \;\; F_t^2(\alpha^2) = \Delta T^2 = \langle \textit{potential time - energy per } \alpha^2 \textit{ unit} \rangle$$

Einstein's Gravitational Field Tensor

$$G_{\mu\upsilon} + \Lambda g_{\mu\upsilon} = \frac{8\pi G}{c^4} T_{\mu\upsilon};$$

$(G_{\mu\upsilon} = \text{Einstein's curvature tensor})$; $(\Lambda = \text{Einstein's cosmological constant})$

$(g_{\mu\upsilon} = \text{Einstein's metric tensor})$; $(T_{\mu\upsilon} = \text{Einstein's stress energy tensor.})$

$(G = \text{Newton's gravitational constant})$; $(c = \text{speed of light in a vacuum})$

Vacuous Space has Proved to be the Quantum Squared, not a Cartesian Solid, Giving a New Value to Einstein's Cosmological Constant

$$\left\langle \begin{matrix} \textbf{\textit{quantum}}\textit{ -squared} \\ \textbf{\textit{vacuum time force}} \end{matrix} \right\rangle = \Lambda = \left\langle \begin{matrix} \textit{Quantum - squared potential time enegy} \\ \textbf{\textit{per}} \text{ meter - squared of vacuum} \end{matrix} \right\rangle = \left(\frac{1}{\alpha^2}\right)\Delta T^2 = \frac{1}{c^2}$$

$$(1/\alpha)^2 = \langle \text{number of "}\alpha^2\text{" units in a "meter}^2\text{"} \rangle$$

$$(1/\alpha)^2 \Delta T^2 = \langle \textit{potential time - energy} \text{ per "m}^2\text{" vacuum} \rangle = F_{time}^2.$$

$$F_{time}^2(\alpha^2) = \Delta T^2; \quad F_{time}^2 = \frac{\Delta T^2}{\alpha^2} = \left(\frac{1}{c}\right)^2; \; c = \frac{\alpha}{\Delta T}; \quad \Lambda = \left(\frac{1}{\alpha^2}\right)\Delta T^2 = \frac{1}{c^2}$$

---

[26] Not only does the graph of redshift to distance for the quantum-curvature model explain more of the variance in the data table than does the graph of Hubble's recession-velocity formula, but Hubble's original formula—inductively concluded from the data table— was revised downward by 90% after his death to accommodate a greater age of the universe. The revision occurred outside data confirmation and in direct conflict with Hubble's original data. See: *The Quantum Dimension;* *"The Quantum Curvature of Space vs. An Expanding Universe. Comparisons by Hubble's original redshift data ."* p. 94. Op. cit.

[27] *Four-Dimensional Atomic Structure:;* L. Dawson, Paradigm Publishing, 2013. See: Tab 6 *"The Derivations of the Alpha Space, the Wave-Phase Time Constant and Planck's Constant from Dawson's Tensor"*

The Heaviside Formulation for the Permeability/Permittivity of Vacuum[28]

Oliver Heaviside proposed a formula for the impact of the two known electrodynamic fields upon vacuum as a simplification of the Maxwell field equations. The first, the electromagnetic field which projects newtons of force as a function of an electrical current measured in amperes (as with an electromagnetic), Heaviside called "the magnetic permeability of vacuum." The second field, the capacitance field which stores energy which is discharges over time (as with an electrical capacitor), Heaviside called the "the electric Permittivity of vacuum." By his formula, "magnetic permeability" *times* "electric permittivity" *equals* "the inverse of the speed of light squared in a vacuum."

HEAVISIDE' S ELECTRODYNAMIC PENETRATION OF VACUUM

$$\mu_0 = magnetic\ permeability = 4\pi\left(10^{-7}\right)\frac{Newtons}{(amp)^2}; \qquad \varepsilon_0 = \ electric\ permittivity;$$

$$\mu_0\varepsilon_0 = \frac{1}{c^2} = 1.1126500561e - 17\ \frac{\sec.^2}{meter^2}; \qquad c = speed\ of\ light;$$

$$\varepsilon_0 = 8.8541878176e - 12\ \frac{Farads}{meter^2}; \ \left(\text{as calculated from Heaviside's value for "}\mu_0\text{"}\right)$$

$$\mu_0\varepsilon_0 = 4\pi\left(10^{-7}\right)\frac{Newtons}{(amp)^2}\left(8.8541878176e - 12\ \frac{Farads}{meters^2}\right) = 1.1126500561e - 17\ \frac{\sec.^2}{meter^2}$$

Heaviside's Formula for the Electric Permittivity and Magnetic Permeability of Vacuum is the Exact Equivalent of the Quantum-Squared Cosmological Constant for Vacuum

Heaviside's formula for the interface of electrodynamic fields with vacuum as equaling "$1/c^2$" is the exact equivalent of the quantum-dimensional force sustaining vacuum. The reserved time energy sustaining the alpha squared vacuole is the square of reserved time separating the quantum end-points which establishes the alpha space of separation. The force sustaining the separation is the summation of all the reserved time energies in a meter squared of alpha squared units. It is also equal to "$1/c^2$":

$$\alpha = \{fundamental\ quantum\} = 0.50216243346e - 15\ meters$$ [29]

$$\Delta T = \{reserved\ time\ across\ alpha\ space\} = 1.6750335776e - 24\ \sec.$$

$$\Delta T^2 = \left\{potential\ time\ energy\ for\ a\ single\ "\alpha^2"\ unit\ of\ vacuum\right\};$$

$$F_t^2\left(\alpha^2\right) = E = \Delta T^2 \quad \left("Force"\ times\ "distance"\ equals\ "energy"\right)$$

$$\frac{1}{\alpha^2} = \left\{number\ of\ "\alpha^2"\ units\ in\ a\ square\ meter\right\} = 3.965624231e30\left(\frac{units\ \alpha^2}{meter^2}\right);$$

$$F_{time}^2 = \left(\frac{1}{\alpha^2}\right)\Delta T^2 \quad \left\langle \begin{array}{l} Time - force\ squared\ equals\ number\ of\ "a^2" \\ units\ per\ meter\ squared\ \text{times}\ energy\ per\ unit \end{array} \right\rangle$$

$$F_t^2 = \left(\frac{1}{\alpha^2}\right)\Delta T^2 = \frac{1}{c^2} = 1.1126500561e - 17\left(\frac{\sec.^2}{meter^2}\right) = \mu_0\varepsilon_0 = \Lambda$$

---

[28] Standard SI formula in physics. SEE Heaviside, Oliver; *Electromagnetic Theory*, Vols. I, II,. and. III. Reprint. . New York:. Dover,. 1950.

[29] Derived from Quantum Atomic Model and Dawson's Tensor. See: Four *Dimensional Atomic Structure.* Tab 6. Op. cit.

THE QUANTUM-DIMENSIONAL TRANSFORMATION Of HEAVISIDE'S "ELECTRIC PERMITTIVITY " UNIT OF MEASURE

Heaviside's "electric permittivity" is a capacitance field unit of measure for vacuum in "Farads/ m2 Capacitance field strengths are given in "Farads" which are measured by the amount of time the field-stored energy takes to discharge (capacitance *times* resistance *equals* time). Time is a function of "charge" which is defined as amperes *times* time. One coulomb of charge *equals* one amp of current flow *times* one second of time.

By the Heaviside formula, however, this capacitance to store energy, as measured in Farads, is a characteristic of vacuum, not of a capacitor in an electrical circuit. It is the "permitted" energy stored in a meter-squared of free space (vacuum). The "meter squared" makes the capacitance field definition coherent with the field definition for an the electromagnetic field projected by a current flow. The standard definition of an amp is the following:

$$\langle \text{the geometric definition of amperage} \rangle = \frac{2(10^{-7})\,Newtons}{meter^2(\text{of separation})}(length)$$ [30]

The "permeability" of vacuum to Newtons of electromagnetic force must be measured in a "meter squared field of vacuum," so the "permittivity" of vacuum to the capacitance storage of energy must also be measured as a "meter squared[31] " unit of the vacuum. The energy stored in the capacitance field is that provided by the vacuum itself which, by Heaviside's formula, is a mathematical function to the quantum-squared vacuum time force:

$$\frac{Farad}{m^2} = \frac{Capacitance}{m^2} = \frac{Charge}{voltage(m^2)}; \quad \frac{(Charge)(Voltage)}{m^2} = \frac{Energy}{m^2} = \frac{\Delta T^2}{\alpha^2} = \frac{F^2_{time}}{m^2}$$

$$Charge = amps(\text{sec.}); \quad \text{Voltage} = \frac{Energy}{Charge} = \frac{F^2_{time}}{amps(\text{sec.})}$$

$$\frac{Farad}{m^2} = \frac{Charge}{voltage(m^2)} = \frac{amps(\text{sec.})}{(m^2)F^2_{time}/amps(\text{sec.})} = \frac{amps^2(\text{sec.}^2)}{F^2_{time}(m^2)}$$

APPLIED TO HEAVISIDE'S ELECTRODYNAMICS "PERMEABILITY/ PERMITTIVITY" OF VACUUM

$$1.1126500561e\text{-}17\ \frac{\text{sec.}^2}{meter^2} = 4\pi(10^{-7})\frac{Newtons}{(amp)^2}\left(8.8541878176e\text{-}12\ \frac{Farads}{meters^2}\right)$$

$$\frac{\text{sec.}^2}{meter^2} = \frac{Newtons}{(amp)^2}\left(\frac{Farads}{meters^2}\right); \quad \left\langle \begin{matrix}\text{factor out the numeric values to}\\ \text{remainder the units of measure}\end{matrix}\right\rangle$$

$$\frac{\text{sec.}^2}{meter^2} = \frac{Newton}{(amp)^2}\left(\frac{amp^2(\text{sec}^2)}{F^2_{time}(meter^2)}\right); \quad \left\langle \begin{matrix}\text{substituting the quantum - squared}\\ \text{time-force value for "permittivity."}\end{matrix}\right\rangle$$

$$\frac{\text{sec.}^2}{meter^2} = \frac{Newton}{F^2_{time}}\left(\frac{\text{sec.}^2}{meter^2}\right)$$

$$\mathbf{Newton = F^2_{time} = kg\frac{meter}{sec.^2}} \quad \left\langle \begin{matrix}\textbf{The quantum - dimensional time-force squared}\\ \textbf{seamlessly integrates with the Newton force unit}\end{matrix}\right\rangle$$

---

[30] Two wires carrying current are separated by a meter of distance for one meter of length making a field of a meter squared providing newtons of force equal to the current. Multiplying the length of wire linearly multiplies the force.

[31] The SI unit-value of "F/m" is wrong. A capacitance field can never be expressed linearly. "F/m2" is correct.

The Math Logic of the Seamless Integration of the Time-Force Squared with the Standard "Newton" of Force

The Heaviside formula shows the penetrability of vacuum by electric and magnetic fields. His magnetic “permeability” is given in “Newtons per amps squared.” His electric field “permittivity” is given in “Farads per meter (squared).” These penetration field values are set equal to the speed of light squared which we have elsewhere shown to be equal to the “time-force squared” which establishes quantum vacuum[32]. A variance in time establishes vacuum by enforcing a space of separation between the incompatible time values.

Because the Heaviside formula incorporates both standard “Newtons” of force and the “time-force squared,” which establishes vacuum, it can be used to calculate an equivalence between the two forms of force. The energy stored in the capacitance field of the Heaviside electric field component can be assigned by the time energy establishing vacuous space. This time energy assignment can then be used to provide a voltage value to the field by conventional electronics mathematics. The Heaviside electric field “permittivity” unit of measure can be rewritten, incorporating the time-energy assignment of capacitance field voltage. “Time-force squared” becomes a component of the Heaviside “permittivity” unit of measure. Factoring out the numeric component of the Heaviside equation allows a direct equality between a “Newton” of force and the time-force squared.

The Quantum Dimensional Mathematics Governing the Establishment of Vacuous Space by Reserved Time Energy

$\Delta T = \langle \text{The potential time-energy variance. Two proximate time points which are ahead or behind one another.} \rangle$

$\alpha = \langle \text{The space of separation enforced between the two incompatible time points.} \rangle$

$\frac{\alpha}{\Delta T} = c = \langle \text{The limit of the speed of light. Time would not exist for greater velocities across the } \alpha \text{ space.} \rangle$

$F_{time} = \langle \text{The force separating the two incompatible time values.} \rangle$; $\quad F_{time}(\alpha) = \Delta T; \quad F_{time} = \Delta T / \alpha = 1/c$

$$\Delta T^2 = \left\langle \begin{matrix} \text{The potential time energy established by the} \\ \text{time variance. The square of the time variance} \\ \text{produces the volume of vacuum.} \end{matrix} \right\rangle; \quad F_{time}^2 = \left\langle \begin{matrix} \text{The time-force squared which} \\ \text{implements volume by "kinking"} \\ \text{Euclidean distances into curvature.} \end{matrix} \right\rangle$$

$$\alpha = 0.50216243346e-15 \ meters; \qquad \Delta T = 1.6749606744e-24 \ \Delta\text{sec}.$$

The Seamless Integration of the “Time-Force Squared” (Cosmological Constant) with Newtons of Force Provides a Direct Interface Between Quantum-Squared Vacuum and Newtonian Gravitational Mechanics

The Newtonian Gravitational equation is the following:

$$F = G\frac{m_1 m_2}{r^2}; \qquad G \approx \langle Gravitational\ constant \rangle \cong 6.67384e\text{-}11\ Newtons \frac{meter^2}{kg^2}$$

Since we are not sophisticated enough to make a “Gordian knot” of Newton’s gravitational constant by increasing the complexity of Newton’s original unit of measure, we will stick with the original

[32] *The Theory of Time-Enforced, Four Dimensional Space;* The Quantum Dimension, p.112. Op. cit.

"m^2/kg^2 Newtons." By seamless integration with time-force squared the constant can be written:

$$G = \langle Gravitational\ Constant \rangle \cong 6.673843e-11\ \left(F_{time}^2\right)\frac{meter^2}{kg^2}$$

The seamless integration of the "time-force squared" cosmological constant with a Newton of force renders Isaac Newton's gravitational equation to be the following:

$$Gravitational\ Force \cong G\frac{m_1(kg)\times m_2(kg)}{r^2\left(meter^2\right)};\quad G \cong 6.673843e-11\ \left(F_{time}^2\right)\frac{meter^2}{kg^2}$$

$$Gravitational\ Force \cong \left(F_{time}^2\right)\left[6.673843e-11\left(m_1^{numeric}\right)\left(m_2^{numeric}\right)\frac{1}{r_{numeric}^2}\right]$$

The force of gravity established between two proximate masses is a function of the time-force expanding space , the numeric values (in kilograms) of the two masses , and the numeric value of the square of the distance of separation (in meters).

The force of gravity is shown to be a mathematical function of the expansion of vacuous space by reserved or potential time energy. This can be proved by quantum dimensional mathematics which can identify the source of gravitational attraction and which can derive Newton's gravitational constant using quantum-dimensional geometry.

The Quantum-Dimensional Theory of Gravity and its Relationship to Newtonian Gravitational Mechanics

The primary thesis of quantum-dimensional gravitational theory is that mass must expand the surrounding vacuous space by a mathematically determined amount. The amount of expansion is a function of mass as the derivative of four-dimensional space and the algebraic transformation which governs the conversion of strict Euclidean space to quantum space.

The forceful expansion of quantum-squared vacuous space by mass is mathematically required to contract back upon the mass. This provides a counter-force relative the expansion force. A spacial-tension field is thus produced. If the spacial-tension fields of two masses are combined, the forces of expansion along the line of opposition between the two masses are neutralized, remaindering only the force of contraction. The unopposed forces of contraction attempts to combine the two masses into a single mass providing a single field. This is the actual source of gravitational attraction.

The force of mass-expanded vacuum resists expansion by attempting to increase the mass's radius of maximum density by an exact amount. This increase is a constant which must be multiplied by radius size to determine total resistance force. This mass-expansion constant is provided for by the vacuum-expansion constant.

This mass-expansion constant *times* the radius supplies the total amount of force applied back against the mass. Since the amount of mass is a mathematical function of the radius, total force is a constant function *times* the mass. Newton's gravitational constant also delivers a total force value as the mass *times* a function of the constant. That function is division by the distance squared between the two oppositional masses.

The expansion force in question is the time-force squared which has been shown to seamlessly integrate with Newtons of force. However, this mass-expanding force cannot actually increase the size of the mass. Prevented from increasing the mass, the attempted mass-expanding time-force is directly converted to Newton's of acceleration per kilograms (squared) of mass. The force in question is only expressed when two masses are in gravitational opposition to one another and, therefore, total force results from the multiplication of the masses of two bodies (rendering a "kilogram squared" unit). We will see how this spacial back pressure against the mass as an "expansion constant" calculates to become a component of Newton's gravitational constant.

MASS AS THE DERIVATIVE OF A FOUR-DIMENSIONAL UNIT OF MEASURE

We have seen that the derivative of an Euclidean solid is its quantum value which is vacuous volume or what science has designated as physical "vacuum:"

$$\textit{quantum value of a solid} = D(x^3) = 3x^2$$ *three planes aligned along three Cartesian axii.*

However, we have also demonstrated that such "strict Euclidean vacuum" is impossible because the floating point of origin for the graph would produce a vacuum which is actually a Cartesian solid. Three Cartesian axii cannot define vacuum since such vacuous volume must be defined from a fixed point of origin which the Cartesian graph cannot provide. Only the graph of the quantum squared provides a fixed point of origin. We have proved that vacuum is quantum-squared by reevaluating the Eddington/Dyson data which shows an horizonal quantum-squared vacuum lensing bias— the very same Eddington/Dyson data which had initially confirmed Einstein's formula for the gravitational curvature of vacuum by mass. The quantum-squared unit of vacuum is "lean" in relationship to a strict Euclidean definition of the vacuum unit. Quantum squared volume is only "1/ 3" of what it would be if it were strictly Euclidean in definition.

Area of Quantum-Squared Vacuum is 1/ 3 the Euclidean Definition

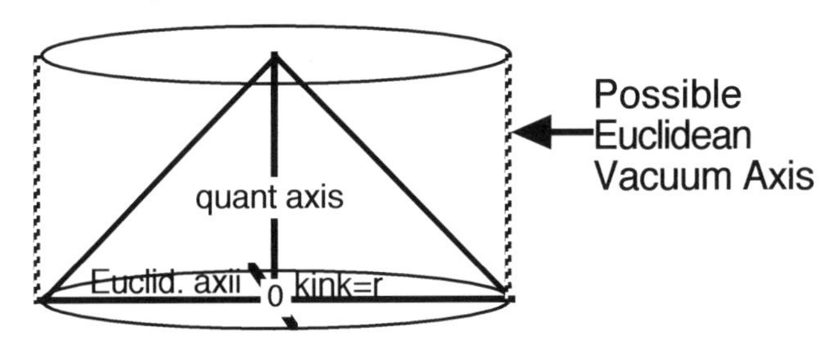

$$\textit{Volume Euclidean Definition} = \pi r^2(\textit{quantum axis})$$

$$\textit{Volume Quantum Squared} = \frac{\pi r^2(\textit{quantum axis})}{3}$$

The Quantum Derivative of Four-Dimensional Euclidean Space is a Unit of Mass which Expands the Surrounding Vacuum.

Mass is four dimensional. Specifically, mass is the quantum derivative of four dimensional space. This produces a three dimensional Euclidean solid with a force projection along the fourth quantum dimension; along the quantum axis not included in the mass's definition of volume. As with the derivative of an Euclidean solid which produces a vacuous volume, so the derivative of four-dimensional space cannot provide a strict Euclidean solution. The solution must be quantum-dimensional. This quantum-dimensional formulation of mass can be proved by the capacity of the model to exactly derive Newton's gravitational constant. We begin with the strict Euclidean derivative of a four-dimensional unit of measure which renders its quantum value. That derivative is the following:

$$\textit{quantum value} = D(x^4) = 4x^3$$

There are not four axii along which Euclidean volume (defined as "x^3") can be arrayed such that they can supply a vacuum of separation between them. The only "rational geometric solution" is the projection of force by the solid along the quantum axis *times* "4;" along the quantum geometric axis which only exists in vacuum and which is not contained in the mass's definition of volume.

The fourth quantum axis exists as a component of the quantum-squared which establishes the volume of vacuum by the "kinking[33] " of an Euclidean axis into curvature. We have demonstrated that the quantum dimension exists externally to the solid and must be graphed as the "quantum squared." Vacuous volume or "vacuum" cannot be graphed using Cartesian coordinates.

For "4" quantum-axii to exist outside a solid, vacuum's quantum-squared must be squared:

$$(\textit{the } 2 \text{ "}Q^2\text{" } \textit{axii})^2 = 4 \textit{ axii}$$

---

[33] "1+1 dimensional kink" *Solitons* by Sascha Vongehr, 1997; physics.usc.edu/~vongehr/solitons_html

By the principle that the derivative of Euclidean space produces its quantum value, the quantum value of a four-dimensional unit of space is an Euclidean solid arrayed on four axii. Only the quantum-squared unit of measure which establishes vacuous volume outside of the solid is available to be multiplied by the solid.

Therefore the derivative of four dimensional space is a geometric solid which is arrayed along the square of the quantum square. The quantum value of four dimensional space is a geometric solid which is arrayed along the square of the two quantum-squared axii :

$$D\left(x^4\right) = (\textit{the 2 axii of vacuum})^2 x^3 = 2^2 x^3 = 4\,x^3$$

The Four-Dimensional Derivative reverses the *Dimensional Authority* of the Quantum Dimension over the Euclidean Dimension

Quantum vacuum must be graphed as the quantum squared which interfaces an Euclidean axis with the quantum dimension from an external quantum point (see illustration on page 2). This intersection from the quantum point imposes a unit of distance upon the Euclidean axis. The Euclidean unit of measure is modified by the quantum dimension because the Euclidean distance is "kinked" into curvature by quantum force. This "kinking" produces volume for and by the quantum-squared unit. Thus, the Euclidean dimension is modified by the quantum dimension because Euclidean distances are kinked into curvature. I will term this control of the Euclidean dimension by the quantum as *"dimensional authority."*

The quantum value of a four-dimensional unit of space is an Euclidean-solid derivative which multiplies or increases vacuum by arraying it along 4 axii which are unique to vacuum. This multiplication is accomplished by a reversal of the *dimensional authority* which the quantum dimension holds over the Euclidean dimensions within the quantum-squared graph of vacuum. The quantum dimension can no longer impose a curved unit of measure upon the three Cartesian axii defining the solid as it does the Euclidean unit of distance within quantum squared vacuum.

Rather, the solid supplants quantum space along all radials and imposes a new quantum definition on surrounding quantum vacuum. The value of "x" for the derivative of the four-dimensional unit supplies a radial value by the following:

$$D\left(x^4\right) = 4x^3; \qquad x^3 = \frac{4\pi r^3}{3} \qquad \{\textit{The volume of a sphere}\}$$

The solid must be spherical because supplanted quantum vacuum provides a counter force in all directions along the surface of the solid. The supplanted quantum-squared "vacuoles" are all sustained by force which "push back" against their supplantation by the solid. The sphere is the geometric form which provides equal counter force resistance across the surface, as in the case of a "bubble."

The Radius of the Solid imposes a new Quantum Value upon the Surrounding Vacuum, thus establishing *Dimensional Authority* over the Quantum Dimension

Since the solid supplants quantum vacuum for the volume it occupies, the solid establishes a new quantum value for the surrounding vacuum equal to the following:

$$Q = (\textit{radius in meters} = r); \quad \alpha = (\textit{fundamental quantum}) \cong 0.50216\,\left(10^{-15}\right)\textit{meters}; \quad \frac{r}{\alpha} = (\textit{alpha spaces in "r"})$$ [34]

There must be an Algebraic Translation for the Multiplication of a Quantum under Euclidean *Dimensional Authority*

Unlike the *quantum dimensional authority* of vacuum, the *Euclidean dimensional authority* of the solid does not supply a one-to-one multiplication of the dependent dimensional unit. For quantum-squared vacuum we know exactly how many Euclidean units there are per quantum. There are exactly one Euclidean unit per quantum which is kinked into curvature to provide a curved value of "$\alpha\pi/2 \cong 1.5708\alpha$." Multiplying the quantum under *quantum dimensional authority*

[34] See *Four Dimensional Atomic Structure;* Paradigmphysics, 2013; Tab 6, *"The Derivation of the Alpha Space....... from Dawson's Tensor"* for the calculated value of the alpha space in meters.

renders a one-to-one algebraic translation with dependent Euclidean units.

However, multiplying the quantum under *Euclidean dimensional authority* supplies no direct algebraic translation. We simply do not know how many quantum units there are in each Euclidean unit. The Euclidean unit is composed of a continuum of points. The quantum unit is composed of only two points. By definition, one could fit an infinite number of quantums into a single Euclidean unit of measure since the Euclidean unit is infinitely divisible and each division supplies a new quantum. In what sense can we multiply an Euclidean unit to render a number of quantums? I propose that the multiplication of an Euclidean unit simultaneously multiplies the number of quantums by dividing the Euclidean unit for each value of the multiple "n," which renders the following algebraic translation :

Proposed Algebraic Translation[35] for the Multiplication of the Quantum by a Solid with Euclidean *Dimensional Authority*

$$n\frac{x^3}{2^n}=(Q=r); \qquad n\left(x^3\right)=\left(2^n\right)(Q=r); \qquad \text{r}=\langle \textit{Mass radius of maximum density} \rangle$$

The Geometric Solid Must Forcibly Expand Surrounding Vacuum

The derivative of a four-dimensional unit of space is a solid which is arrayed along four quantum axii in the surrounding vacuum. The solid has *dimensional authority* in that it imposes a new quantum value upon vacuum; a quantum-value which is a mathematical function of the radius of the mass. This new quantum value is established by a quantum force equal to the radius *divided by* the alpha space *times* the force establishing a single alpha space[36] . The solid must multiply vacuum by, at least, the factor of "4" which is the factor established by the derivative. The forcible expansion of vacuum by the solid is accomplished by multiplying the new radius-determined quantum by a factor which is established by the mass.

However, an Euclidean measure with *dimensional authority* cannot directly multiply a dependent quantum measure. The *Euclidean dimensional authority* imposes an algebraic translation formula upon the quantum for multiplication (as noted above). This conversion formula places additional mathematical constraints upon the mass itself. As the multiple "n" becomes larger, the force of expanding quantum space establishes increasing "back pressure" against the radius of the mass and against the mass's volume:

The Constraint on the Mass's Radius by the Algebraic Translation of "n" Causes a Counter Force which is a Component of the Force of Gravity

$$x^3=4\frac{\pi r^3}{3}=4.1887902048r^3; \quad r=\frac{x}{\sqrt[3]{4.1887902048}} \qquad \left\{\begin{matrix}\text{"r" is a direct function}\\ \text{of the unit of volume.}\end{matrix}\right\}$$

$$\frac{nx^3}{2^n}=(Q=r); \qquad Q=quantum$$

$$\frac{n\left(4.1887902048r^3\right)}{2^n}=Q; \quad \frac{n\left(4.1887902048r^2\right)}{2^n}=\frac{Q}{r} \qquad \left\{\begin{matrix}\text{Quantum to radius}\\ \text{ratio should be "1."}\end{matrix}\right\}$$

$$r^2=\left(\frac{2^n}{4.1887902048n}\right)\frac{Q}{r}; \quad r=\sqrt{\left(\frac{2^n}{4.1887902048n}\right)\frac{Q}{r}} \qquad \left\{\begin{matrix}\text{"r" is direct function of}\\ \text{"n" and the "Q/r" ratio.}\end{matrix}\right\}$$

By the above formula, it can be shown that the value of "r" is dependent upon the value of "n." As "n" increases, expanding quantum space attempts to increase "r" and thus the volume of the

[35] This is an inductively concluded algebraic translation which is supplied real-world confirmation in the measured relationship between the distance to the sun's nearest neighbor, Proxima Centauri, and the sun's radius of maximum density.

[36] Equal to the square root of time-force squared, or the square root of one Newton of force per meter of vacuum.

solid. However, because of the *Euclidean dimensional authority* of the solid as well as gravity, the radius of the solid cannot be increased by the counter force of expanded quantum vacuum. This produces a contraction force against the force of expansion and a tension.

The Contraction Force to Vacuum Expansion is a Factor in the Force of Gravity

For any value of "n" the quantum vacuum attempts to impose a value upon the radius of the solid. However, the value of the solid's radius "r" is dependent upon the cube root of "n."

$$\{\Delta \text{r from n.}\} = \sqrt[3]{n}(r);\ \ \{\text{"r" expansion by quantum backpressure}\} = \sqrt{\frac{2^n}{4.1887902048n}\left(\frac{Q}{r}=1\right)}$$

Actual change in "r" value, as determined by "n," is the cube root of "n." An alternative "r" value is attempted by quantum back pressure. Alternative attempts greater than actual create tension.

For any value "n" greater than "5.26," quantum-attempted radius multiplication will exceed the solid's actual change in radius via "n."

$$\sqrt[3]{n}(r) \le \left[r = \sqrt{\frac{2^n}{4.1887902048n}\left(\frac{Q}{r}=1\right)} \right] \ \ \langle when \text{ "n} \ge 5.26\text{"}\rangle$$

If "n " is in excess of 5.26, then quantum vacuum will attempt a radial increase against the solid's *dimensional authority.* Since increasing the solid's radius is impossible, the quantum attempt resides as unresolved tension. This unresolved tension provides a "non-decaying moment of force."

Newton's Gravitational Constant as a Non-Decaying Moment of Force

The gravitational constant is a moment of constituent force which does not decay over time. A constituent force is a force which is provided as a component of a single energy system. The most common example of a constituent force is that provided by a tensioned vibrating string.

Tensioned stings vibrates at constant frequencies regardless of the amplitude of vibrations. However, all amplitudes of vibration decay to the "0" amplitude wave which is the tension constant for the string[37]. The tension constant provides decaying moments of constituent force until a "0" moment of force is reached with a stilled string.

In contrast to the tensioned string and its decaying constituent moment of force, Newton's gravitational moment of force does not decay. In the case of the string, the energy gained as enforced motion is surrendered to re-stretching the sting with a slight energy loss. Thus the moment of force decays over time. With Newton's moment of constituent force (his gravitational constant), however, no energy is lost in the enforcement of motion by the force. None of the energy supplied by the moment of force is surrendered to the gravitational field and the moment of force does not decay. As the objects close, a higher force value is imposed by the unchanged constituent moment of force with no energy lost to the gravitational field. This unique feature of a non-decaying moment of force for a gravitational field make such fields open-energy systems[38] which create energy. Proof of this is supplied by the exact derivation of Newton's gravitational constant using the quantum-dimensional model of mass and its expansion of surrounding vacuum.

Newton's Gravitational Equation and his Non-Decaying Moment of Force

$$F = G\frac{m_1 m_2}{d^2};\ \ G = \{Gravitational\ Constant = Non\text{-}Decaying\ Moment\ of\ Force\}$$

$$G \cong 6.67384e\text{-}11\ Newtons\left(or\ F^2_{time}\right)\frac{meter^2}{kg^2} \quad \langle \textit{Current SI value} \rangle$$

[37] See Tab 2, *"The Failure of the Schrödinger Model of Electron Orbitals;"* p.p. 4-5 in *Four-Dimensional Atomic Structure.* for the graph of the tensioned string's decay to the "0" amplitude tension constant. Op. cit.

[38] *"THE QUANTUM MECHANICS OF A GRAVITATIONAL OPEN ENERGY SYSTEM ;"* Dawson, Lawrence, SRNRL. *http://www.paradigmphysics.com/gravity_open_energy.pdf*

The Derivation of Newton's Non-Decaying Moment of Force for the Gravitation Field

We have proved that the algebraic translation by which mass expands surrounding quantum vacuum also creates a back pressure which attempts to increase the radius of the mass, if "n>5.26." Since mass is the derivative and quantum value of four-dimensional space, It must have an "n" value of at least "4." Mass is four-dimensional and the only way its three dimensional volume can be arrayed along the fourth quantum dimension is multiplying its volume to extend quantum vacuum.

$$n=(\textit{derivative value})(\textit{number of radii in diameter})^3=4(2)^3=32\geq 5.26$$

The quantum must be equal to the radius, not the linear value of the whole mass (the diameter). This requires multiplication by "8" (2 cubed) to geometrically achieve the algebraic translation.

Quantum back pressure attempts to expand the radius "5660.576145 *times*" (n=32) while the actual four dimensional value of the radius is "3.1748 *times*."

$$\sqrt[3]{32}(r)=3.1748021039(r_{\text{actual}});\quad r_{\text{Q expansion}}=\sqrt{\frac{2^{32}}{(4.1887902048)(32)}}=5660.576145$$

"N" multiplies a mass's volume, not its "weight." Mass is characterized by "density," which is its ratio of "weight" to volume. All mass is four dimensional with a three-dimensional volume which is the quantum derivative of four-dimensional space arrayed along the four axii. The four axii only exist as the three axii of volume and a projection to the quantum axis (defining vacuum) outside the three axii of solid volume. The four-dimensional definition of mass requires that its volume be multiplied and, therefore, that its density is less than maximum. The radius of maximum density "r" defines the quantum. The actual density *must be a whole number multiple of "r."* From the above equality, it can be seen that the "n multiple of 32" does not meet this condition. Counter force tensions must reduce "n" to the following:

$$(32)x^3=2^{32}(Q=r);\quad x^3=\frac{2^{32}(Q=r)}{32}=2^{27}(Q=r);\quad n=27;\quad \sqrt[3]{27}(r)=3(r_{\text{actual}})$$

The amount of potential tension produced by mass's "2^{27}(r)" expansion of external vacuum can be calculated using the attempted radius expansion by quantum back pressure[40] ."

$$\{\text{attempted "}\Delta\text{r" expansion by quantum back pressure}\}=\sqrt{\frac{2^{27}}{(4.18879)(27)}}$$

$$\Delta r^2=\frac{2^{27}}{(4.18879)(27)}=\{\text{maximum distance (squared) across the tension field}\}$$

Quantum attempts to expand the radius and the mass's resistance to this radial expansion produces a "tension field." The "tension field" is defined as the "field of resistance to the expansion of the radius." The "field" consists of the amount of tension between quantum expansion and the radial resistance to expansion when the radius is actually "contracted back to the original." The maximum tension exists at no radial expansion and minimum tension exists at maximum radial expansion or at "r=Δr . " Gravitational attraction is a function of this tension field. The amount vacuum counter-force tension across the field can be directly calculated using the Quantum Open-Energy Integral.

The Universal Quantum Open-Energy Integral for Field Force[41]

$$x=\{\text{distance across field (in quantum units)}\};\quad F_{\text{max}}=\{\text{maximum force of field}\}$$

$$\left(1-\frac{1}{x}\right)F_{\text{max}}=\int_1^x\frac{F_{\text{max}}}{x^2}d(x)$$

The "Δr " can represent the distance of a fall across a field if the mass doesn't completely prevent

[40] See p.19-20.

[41] See "The Universal Quantum Open-Energy Integral for Field Force," in Four Dimensional Atomic Structure, Tab 8; "Quantum Identified Field Generaed Energy and its Application to the Asymmetrical Nuclear Capacitor;" p. 128; Dawson, Lawrence. Op Cit.

any change in its radius, but “allows” vacuum counter-force to increase radial distance to the new quantitized value, then suppresses the quantitized increase in its radius back to “1,” or to the position for which the mass’s radius is unchanged. The force of expansion of vacuum as provided by the mass would remain constant. The total counter force in the fall back to “1” would be expressed by the following:

The Universal Quantum-Open Energy Integral applied to Vacuum Back Pressure

$$\{Total\ Contraction\ Force\ across\ fall\} = \left(1 - \frac{1}{\Delta r}\right)(Vacuum\text{ - }Expansion\ Force)$$

$$= \int_1^{\Delta r} \frac{(Vacuum\text{ - }Expansion\ Force)}{\Delta r^2} d(\Delta r);$$

This allows us to derive a moment of force for the actual attempted expansion of the radius of the mass which is the following:

The Derivative of the Quantum Integral provides a Back-Pressure Moment of Force

$$\{Moment\ of\ Contraction\ Force\ at\ "\Delta r"\} = D\left(\left[1 - \frac{1}{\Delta r}\right](Vacuum\text{ - }Expansion\ Force)\right)$$

$$= \frac{(Vacuum\text{ - }Expansion\ Force)}{\Delta r^2}$$

In turn we can put an actual value on this moment of contraction force using the algebraically translated expansion of quantum vacuum by mass:

The Actual Value of the Back-Pressure Moment of Force

$$\Delta r^2 = \left(\frac{2^n}{4.1887902048n}\right);$$

⟨See page "19"⟩

$$\frac{1}{\Delta r^2} = \frac{4.1887902048n}{2^n}$$

$$\{Moment\ of\ "\Delta r"\ contraction\ force\} = \left(\frac{4.1887902048n}{2^n}\right)(Vacuum\text{ - }Expansion\ Force)$$

By this value we can begin to derive Newton’s “non-decaying moment of force” or his gravitational constant.

The Initial Application of the Back-Pressure Moment of Force to Newton’s Constant

$$G = \begin{Bmatrix} Gravitational\ Constant\ (a\ Non\text{-} \\ Decaying\ Moment\ of\ Force.) \end{Bmatrix} = func.\left(D(1 - \frac{1}{\Delta r})\right)F_{exp.} = func.\left(\frac{1}{\Delta r^2}\right)F_{exp.}$$

$$G = func.\left(\frac{4.1887902048n}{2^n}\right)F_{exp.};\quad F_{exp.} = \{Vacuum\text{ - }Expansion\ Force\}$$

Newton’s Non-Decaying Moment of Force (Gravitational Constant) must be a Mathematical Function of the Moment of Vacuum Back Pressure, not a “One-to-One” Correspondence

There is not a one-to-one correspondence between Newton’s moment of force and the moment of force for expanded-vacuum back pressure against the radius of the mass. This is true because mass expands vacuum along the single quantum axis, but vacuum back pressure attempts to expand mass along all three of mass’s axii. A partial of “G” is the following function:

$$\{partial\ of\ "G"\} = \left(\frac{1}{\Delta r^2}\right)^{3/2} F_{exp.} = \left(\frac{4.1887902048n}{2^n}\right)^{3/2} F_{exp.}$$

The Second Back-Pressure Force

The moment of force from quantum vacuum's attempted expansion of the radius of the mass is only a partial of the gravitational constant because it is modified by a second back pressure. This second back pressure results from the requirement that mass expand vacuum along the quantum axis extending from the radius of mass. However, the quantum axis is not equal to the quantum within the "at-rest" quantum-squared vacuole. Mass cannot expand vacuum by a whole number value of the quantum along the quantum axis.

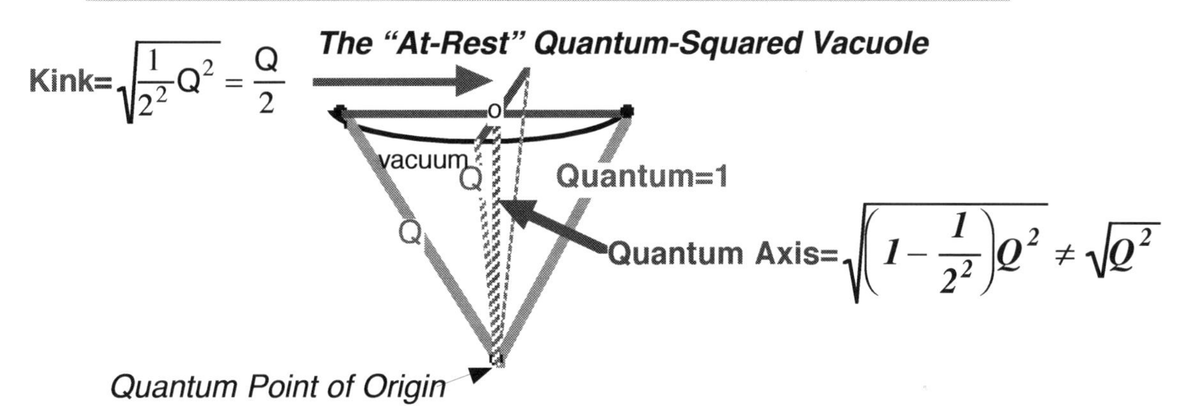

A rational "whole-number" value for the quantum axis can be achieved if the quantum axis (squared) is compressed to "1/ 3" of the quantum-squared. For the "at-rest" vacuole, the quantum axis (squared) is "3/ 4" of the quantum-squared.

$$Q^2 = \{kink\}^2 + \{quantum\text{-}axis\}^2$$

$$\langle at\text{-}rest\ vacuole\rangle: Q^2 = \frac{Q^2}{4} + \frac{3}{4}Q^2; \quad \{quantum\text{-}axis\}^2 = \frac{3}{4}Q^2$$

$$\langle compressed\ quantum\text{-}axis\rangle: Q^2 = \frac{2}{3}Q^2 + \frac{1}{3}Q^2; \quad \{quantum\text{-}axis\}^2 = \frac{1}{3}Q^2$$

Adjusting the vacuole size of the compressed quantum axis by the vacuum-expanding mass produces an exact quantum value for the quantum axis.:

$$(3)Q^2 = (3)\frac{2}{3}Q^2 + (3)\frac{1}{3}Q^2 = 2Q^2 + Q^2; \quad \{quantum\text{-}axis\}^2 = Q^2$$

$$\{quantum\text{-}axis\} = Q$$

Increasing the vacuole quantum by "3 times" and suppressing the axis quantum (squared) to "1/3" produces an exact quantum value. The figure "3" is doubly significant.

Compressing the Quantum-Axis (Squared) to "1/ 3" of the Quantum-Squared Produces Maximum Vacuole Volume and a Counter Force Equal to the Kink Squared

The "kink-squared" is a vector of force for time-force-squared which, as we have proved, seamlessly integrates with mechanical force (*see page 14*).

The "1+1 dimensional kink"[42] is a concept of soliton mathematics. It proposes that a single dimensional line can be forcibly projected into vacuum at 90° to the original line. This forcible projection into a second dimension is called a "kink" and accurately describes what happens when the quantum dimension intersects an Euclidean dimensional line from a single point.

The quantum is defined as the forcible separation of two unlike time values. These offset time values require an exact distance of separation to avoid time incompatibility. A set of quantums is formed by the continuum of points along the Euclidean dimensional line which are shorter than the original quantum distance. These quantums of deficient distance possess an excess of force which

[42] *Solitons,* Vongehr. Op. cit.

must be expressed by kinking the Euclidean line into curvature in order to form quantums at the prescribed distance of separation. A set of quantums established by points along the Euclidean dimensional line are shorter that the original quantum. Forces are equalized by by adding a kink vectors. Compressing the quantum axis, further increases the kink vector and enlarges volume.

Changes in Vacuole Volume by Compressing the Quantum Axis

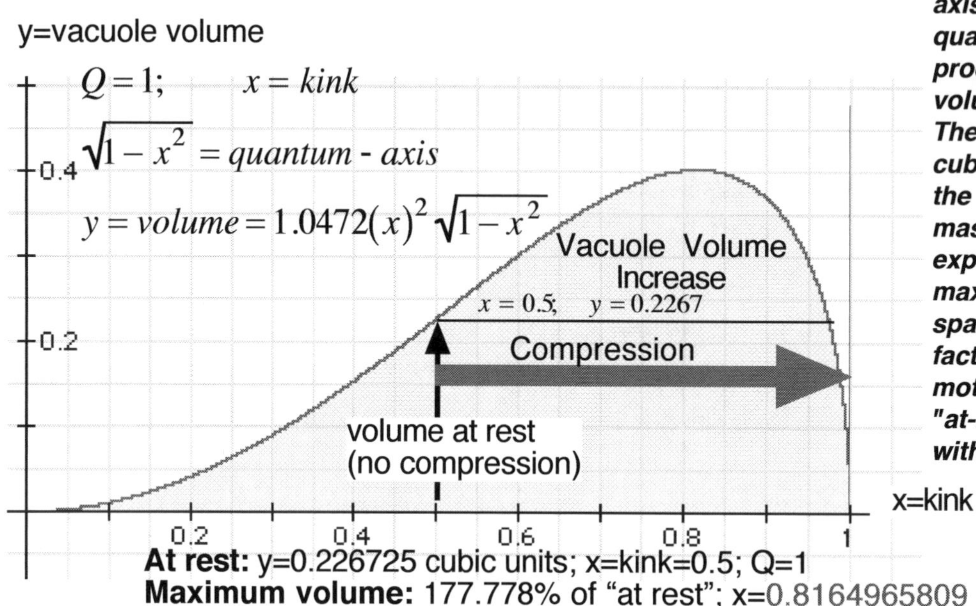

To suppress the quantum axis to "1/3" the vacuole quantum (squared) produces the maximum volume for the vacuole. The number "3" is the cube root of the "n=27" for the expansion of space by mass. That is, the vacuum expanded by mass is at maximum volume for spacial vacuum. It is this factor which allows the motion of mass through "at-rest" vacuous space without resistance.

At rest: y=0.226725 cubic units; x=kink=0.5; Q=1
Maximum volume: 177.778% of "at rest"; x=0.8164965809

Force compresses height which extends "kink" radius because Quantum must be held constant.

$h=\{quantum\ axis\};\quad k=\{kink\}$

$$\text{Volume}=\frac{1}{3}\pi k^2\sqrt{(1-k^2)Q^2};\ \text{let } Q=1$$

$$\text{Volume}=1.0472(k^2)\sqrt{(1-k^2)}$$

The Quantum Vacuole Model

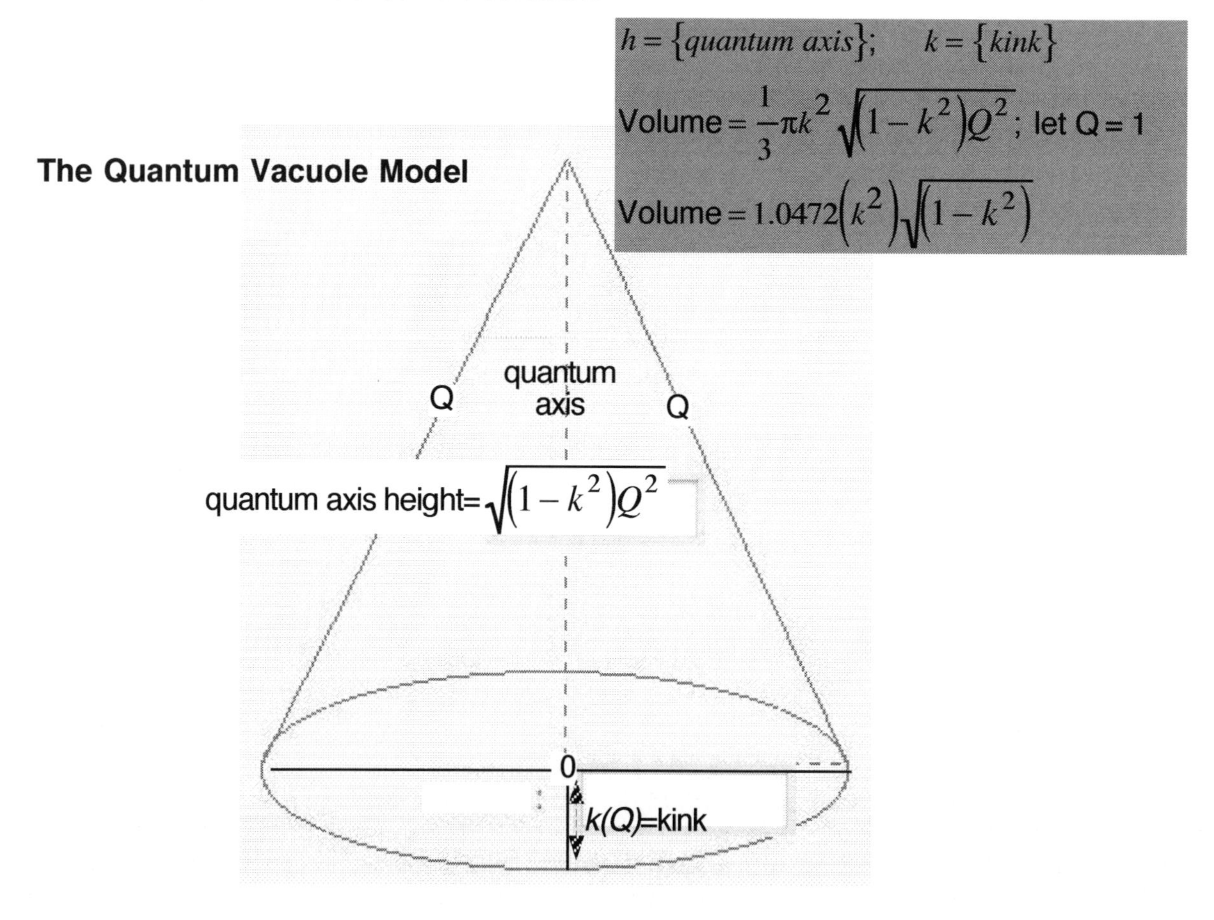

Compressing the Quantum Axis to "1/ 3" the Quantum Squared Maximizes Vacuole Volume as well as Kink Counter-Pressure Efficiency

$$3Q^2 = Q^2 + 2Q^2; \qquad (Quantum\ axis)^2 = \frac{1}{3}Q^2; \qquad (Kink)^2 = \frac{2}{3}Q^2$$

$$Volume = \left(\pi\frac{2}{3}Q^2\right)\left(\sqrt{\frac{1}{3}Q^2}\right)\Big/3 = 0.4030665254Q^3 \ \langle \textit{Maximum vacuole volume} \rangle$$

$$\{At\text{ - }Rest\ Volume\} = \left(\pi\frac{1}{4}Q^2\right)\sqrt{\frac{3}{4}Q^2}\Big/3 = 0.2267249205Q^3$$

$$Kink = 0.8164965809Q \ \langle \textit{Maximum kink pressure from volume contraction} \rangle$$

Kink Reinforcement of "Change in Radius" Back Pressure occurs across the Quantum Axis and Must be Spread into The Volume of Mass

$$\langle \textbf{\textit{Kink reinforcement}} = \kappa\rho \rangle$$

$$4\frac{\pi n\Delta r^2}{3} = 4.1887902048n\Delta r^2 \xleftarrow{"\kappa\rho"\ across\ quantum\ axis} 0.8164965809(Q=1)$$

$$n\Delta r^2 \xleftarrow{"\kappa\rho"\ across\ quantum\ axis} \frac{0.8164965809}{4.1887902048}; \qquad$$ ⟨*See pages 19-20*⟩

Newton's Non-Decaying Moment of Force (Gravitational Constant) is Equal to the Moment of Vacuum Back Pressure *times* Kink Reinforcement

$$x = \{The\ inductively\ concluded\ Euclidean\ "volume\ bias"\ factor\}$$

$$G = \left(D(1-\frac{1}{\Delta r})(\kappa\rho + x)\right)^{3/2} F_{time}^2 = \left(\left(\frac{1}{\Delta r^2}\right)(\kappa\rho + x)\right)^{3/2} F_{time}^2 = \left(\left(\frac{1}{\Delta r^2}\right)(\kappa\rho + x)\right)^{3/2} F_{time}^2$$

$$\boldsymbol{G = \left[\left(\frac{4.1887902048n}{2^n}\right)\left(\frac{0.8164965809}{4.1887902048} + x\right)\right]^{3/2} F_{time}^2;\ F_{time}^2 = 1/c^2 = Newton}$$

Determining the Actual "n" Value for the Quantum-Dimensional Definition of Mass

We have demonstrated that the derivative of an Euclidean unit of measure is always the unit's quantum value and that the derivative of a four-dimensional unit is an Euclidean solid which must be arrayed along four quantum axii. A single quantum axis only exists in the vacuous space outside of the solid. Therefore, the solid must be arrayed four times along the single vacuous quantum axis and this multiplies quantum vacuum.

The quantum value of a four-dimensional unit of measure is an Euclidean solid which multiplies surrounding quantum space along the quantum axis. This expansion has an "n" value of "4."

We have also demonstrated that the multiplication of the solid's radius by an "n" factor of "4" is insufficient to provide the counter force of contraction to vacuous expansion, a contraction which establishes gravity. However, the geometry governing the relationship between the Euclidean solid and the surrounding quantum vacuous space shows that the "n" value is actually "32," not "4." This "n" factor of "32" is sufficient to establish gravitational counter force to the expansion.

The solid linearly supplants quantum space to twice the new value of quantum "r." That is the volume of the solid is eight times that which it would be had it only supplanted the linear value of the new quantum. The diameter is "2r," not the single value "r[43]." Therefore, the solid must multiply the linear quantum by an additional factor of "8" not the factor of "1" had the solid only supplanted a single linear quantum. The new "n" value for the quantum derivative of the four-dimensional unit is the following:

$$4(8)x^3 = 32\frac{4\pi r^3}{3} = 2^{32}(r = Q)$$

Gravity is a Function of Mass but Density Changes by the Cube of Diameter-to-Radius

For the mass to increase its linear value to its radius rather than its diameter, the volume of the solid must increase by the cube of "2":

$$\{maximum\ density\} = \rho_{\max} = \frac{mass}{4\pi r^3/3}; \qquad \{derived\ density\} = \rho_{\text{der}} = \frac{mass}{(4)(2^3)4\pi r^3/3}$$

Because three-dimensional mass is the derivative of four dimensional space, the derivative must expand both the volume of the mass and expand the quantum space surrounding the mass. Because the expansion is a function of the radius (Q=r) and not the diameter, both volume and quantum space must be expanded by an additional "2^3."

The expansion of space must reflect whole number multiples of the quantum radial value. Similarly, the cubed root of the expansion must also reflect a whole number multiple. This is a condition imposed by quantum space.

The cubed root of the quantum expansion value of "2^{32}" does not meet this condition:

$$\sqrt[3]{32} = 3.1748021039; \qquad \sqrt[3]{2^{32}} = 1625.4986772154 \quad \{not\ whole\ numbers\}$$

The force of expansion must contract back to a whole number "n" for which the cubed root of "n" on both sides of the algebraic translation are whole number quantum values. There is a mathematical series which can supply this condition to the quantum side of the equation. If the derivative value of "4" (representing four axii) is reduced to "3" (representing the three axii of volume), then all multiples of "3" for any whole number extensions of vacuum by mass volume produces a "2^n" value with a whole number cubed root.

Geometrically, the value of "n" is always the following:

n=(number of axii which array the solid) (the number of units by which mass's volume extends vacuum)

Thus a mass arrayed along "4" axii with a diameter of "2Q" (radius=Q) has a volume which extends vacuum by "8" units and an "n" value of "32." This is an irrational "n" value for the quantum because it cannot produce a whole-number cubed root for "2^n," nor for "n." This is not the case if we use "3" axii, rather than "4."

Three axii space for which the mass's volume extends vacuum from "1" to "10" times produce A series of whole numbered cubed roots for "2^n." This is a rational quantum-dimensional series for the quantum side of the translation equation in which each succeeding cube root has a whole number value of twice that of its predecessor. The expansion of quantum space must contract to the condition representing a three axis volume. *The expansion of space by a mass must impose an Euclidean set of axii upon the spacial expansion.*

The first ten values of mass-extended vacuum produce a quantum rational "2^n" value with a whole

[43] This supplanting of quantum space by twice the quantum value is empirically proved by the measurement of the alpha space using the electron orbital model. Alpha was calculated to be 1/2 the diameter of the smallest measurable particle, the proton. It was calculated to be equal to the proton's radius. See *Four Dimensional Atomic Structure;* Tab 6, *"The Derivation of the Alpha Space, Planck's Constant and the Wave-Phase Time Constant from Dawson's Tensor"* Op. Cit.

number cube root when multiplied by "3." However, only one of these meets the condition of a perfect translation; that is, has a perfect cube root on both sides of the translation equation. That number is "n=3(9)=27."

Algebraic translation: n(mass)=2^n (quantum vacuum)			
n	*Euclidean side: Cube root of "n"*	*Translation*	*Quantum side: Cube root "2^n"*
3(10)=30	cube root=3.107232506	non-perfect	cube root=1024
3(9)=27	***cube root=3***	***perfect***	***cube root=512***
3(8)=24	cube root=2.8844991406	non-perfect	cube root=256
3(7)=21	cube root=2.7589241764	non-perfect	cube root=128
3(6)=18	cube root=2.6207413942	non-perfect	cube root=64
3(5)=15	cube root=2.4662120743	non-perfect	cube root=32
3(4)=12	cube root=2.2894284851	non-perfect	cube root=16
3(3)=9	cube root=2.0800838231	non-perfect	cube root=8
3(2)=6	cube root=1.8171205928	non-perfect	cube root=4
3(1)=3	cube root=1.4422495703	non-perfect	cube root=2

There is only one of these 10 mass units of vacuum extension which provides a perfect algebraic translation when arrayed along three axii. That number is "9":

$$\{n_{reduced}\} = 3(9) = 27$$

Three spacial axii times nine units of extension.

$$\sqrt[3]{27} = 3; \qquad \sqrt[3]{2^{27}} = 2^9 = 512$$

An "n" of "27" meets Quantum Restrictions on Vacuum Expansion by Mass and Accurately Derives Newton's Gravitational Constant

Quantum-dimensional mathematics defines mass as four dimensional. Mass's affiliation with this extra dimension— the dimension which is not contained in its volume— is as an expansive force along the quantum-dimension. The quantum dimension establishes external vacuum as quantum-squared vacuoles. These quantum squared vacuoles are kinked into volume by the force of reserved time (establishing potential time energy). An analysis of contemporary electrodynamic field equations demonstrated that this quantum-squared time force integrates seamlessly with standard Newtons of force.

The expansion of vacuum along the quantum axis by mass is controlled by an algebraic translation of Euclidean multiples to quantum multiples. This translator establishes that a counter-force from expanded vacuum attempts to increase the radius of the mass which is not possible. However, this counter-force attempt at radial expansion is converted to an attraction force when the vacuum is anchored by a second mass contained within the expansion field. That is, vacuum counter-force is identified as establishing the mathematics of a gravitational system. Newton's gravitational constant is demonstrated to be a function of this attempted radial expansion by quantum space.

The force of gravity is given measurement in the gravitational equation by Newton's gravitational constant which is a non-decaying moment of force. It is multiplied by the product of the two masses and divided by the distance (squared) between the two to give the total moment of force at any one time in an actual gravitational opposition.

The moment of contraction force by mass-expanded vacuum is a mathematical function of the value of "n" in the algebraic translator. Because "n" is a multiplier of mass on the Euclidean side of the translator and a multiplier of a linear quantum on the quantum side of the translator, the mass's radial is actual a cubed value relative to the quantum side's linear value. Therefore, "n" must be a rational cube root on both sides of the translator. That is, it must provide a perfect, whole-number translation for the cube root to both sides of the equality.

The initial value of "n" as calculated by quantum-dimensional mathematics fails to provide this perfect translation of the cube root. Only by reducing "n" from "32" to "27" do we achieve this perfect cube-root translation, as demonstrated above. An "n" value of "27" used in the moment of contraction force equation accurately derives Newton's gravitational constant.

However, the "n=27" value applied to the moment of contraction force must be modified by a second factor imposed by the quantum-squared vacuole. Mass's expansion of vacuum along the quantum axis must be a whole-number multiple of the radial quantum. However, the unit of distance along the quantum axis is not equal to the volume quantum. The unit distance along the quantum axis can only be made equal to the radial quantum by compressing the quantum axis unit (squared) to "1/3" of the quantum squared and multiplying the volume quantum (squared) by "3." This produces a quantum axis unit which is exactly equal to the radial quantum.

This compression of the quantum axis maximizes both vacuole volume and the counter pressure of the kink. The counter-pressure force of the kink reinforces the moment of contraction force supplied by vacuum's attempt to increase the radius of the mass. Using "n=27" in the quantum dimensional equation we get an accurate calculation of Newton's gravitational constant[44] :

Calculating Newton's Gravitational Constant by a "2^{27}" Expansion

$$x = \{\textit{The inductively concluded Euclidean "volume bias" factor*}\} = 0.0003334$$

$$G = \left(D(1 - \frac{1}{\Delta r})(\kappa\rho + x) \right)^{3/2} F_{time}^2 = \left(\left(\frac{1}{\Delta r^2} \right)(\kappa\rho + x) \right)^{3/2} F_{time}^2$$

$$G = \left[\left(\frac{4.1887902048n}{2^n} \right) \left(\frac{0.8164965809}{4.1887902048} + x \right) \right]^{3/2} F_{time}^2$$

$$G = \left[\left(\frac{4.1887902048(27)}{2^{27}} \right) \left(\frac{0.8164965809}{4.1887902048} + 0.0003334 \right) \right]^{3/2} F_{time}^2 = (6.67384e\text{-}11) F_{time}^2$$

$F_{time}^2 = newton$ ⟨see page 14⟩

$$\langle \textbf{SI value of "G"} \rangle = (6.67384e\text{-}11)\ \textbf{newtons} \frac{m^2}{kg^2}$$

$$= \left[\left(\frac{4.1887902048(27)}{2^{27}} \right) \left(\frac{0.8164965809}{4.1887902048} + 0.0003334 \right) \right]^{3/2} F_{time}^2$$

**The "volume bias factor" is produced by the fact that mass expands the volume of vacuum to its maximum. The vacuous space expanded by mass is also maximum volume space. This "pressure vacuum" pushes "at rest" vacuum out of its way and is the reason that that the force establishing vacuum gives no resistance to the motion of matter. It is the equivalent of Einstein's space as a "flexible membrane" which, similarly, is said to remove spacial resistance to the motion of mass.*

[44] The kink reinforcement of the moment of force must incorporate an inductively concluded "volume bias" factor. The existence of such an unknown had confounded the attempts to measure Newton's gravitational constant which is still considered an approximation.

Made in the USA
Middletown, DE
20 July 2021